Emerging Trends in Mechatronics

Series Editor

Aydin Azizi, Oxford, UK

Aydin Azizi
Editor

Smart Diagnostics and Predictive Maintenance

Springer

Editor
Aydin Azizi
School of Engineering, Computing
and Mathematics
Oxford Brookes University
Oxford, UK

ISSN 2731-4855 ISSN 2731-4863 (electronic)
Emerging Trends in Mechatronics
ISBN 978-981-95-3920-8 ISBN 978-981-95-3921-5 (eBook)
https://doi.org/10.1007/978-981-95-3921-5

This Springer imprint is published by the registered company Springer Nature Singapore Pte Ltd.
The registered company address is: 152 Beach Road, #21-01/04 Gateway East, Singapore 189721, Singapore

Preface

This volume, titled *Smart Diagnostics and Predictive Maintenance*, presents a comprehensive collection of advanced research that integrates intelligent control strategies, artificial intelligence, and data-driven methods across energy systems and industrial maintenance. The book begins with a state-of-the-art approach to optimizing grid-connected hybrid PV-battery-supercapacitor systems using machine learning, addressing the growing demand for sustainable and efficient renewable energy solutions. Subsequent chapters explore diagnostic and predictive maintenance methodologies applied to critical industrial components, including reciprocating compressors, electrical submersible pumps, vehicle cylinder heads, and centrifugal pumps.

By combining energy optimization with AI-based condition monitoring and predictive analytics, the chapters in this volume offer both theoretical insights and practical implementations. These contributions aim to bridge the gap between academic research and industrial application, ensuring that solutions are not only innovative but also viable for real-world deployment.

This work will be of interest to researchers, engineers, and professionals working in renewable energy, mechanical systems, process engineering, and artificial intelligence, providing both foundational knowledge and cutting-edge developments.

Oxford, UK — Aydin Azizi

About This Book

The content of this book reflects the convergence of renewable energy optimization and industrial predictive maintenance. It has been structured to present a logical flow from energy system control to equipment fault diagnostics, enabling readers to see the technological synergies between these areas. The opening chapter focuses on intelligent optimization of hybrid PV-battery-supercapacitor systems, offering novel solutions for grid-connected energy management. The remaining chapters address AI-based diagnostics and predictive maintenance for various industrial assets, including compressors, pumps, and engine components.

The aim is to provide an integrated perspective where sustainable energy management and smart maintenance strategies can be developed in tandem. Readers will benefit from experimental data, computational modeling, and applied AI techniques, all presented within the context of solving real-world engineering problems.

Contents

Intelligent Control Strategy Optimization of Grid-Connected Hybrid PV-Battery-Supercapacitor Systems Using Machine Learning

Amer Bashir Omar Aboukhris and Mohammad Karimzadeh Kolamroudi

1 Introduction

1.1 Importance of the Study

During the past decades, global electricity consumption has significantly increased, primarily driven by the immense growth of industry, transportation, and communication means. The energy supply is majorly dependent on fossil fuels such as petroleum. However, this resource is limited and result in global energy shortage in the future. Additionally, the exploitation of these resources results in the emission of highly polluting gases, exacerbating the warming of our planet. It is therefore impossible to continue without an energy-efficient approach.

To address this overall energy crisis and its basic adverse effects, recent developments in the energy sector revolves around promoting alternative energy sources. These new energy resources are renewable in nature and offer solutions to the limitations faced by conventional energy sources. They are inexhaustible, have unlimited availability as they are independent of consumption, and, most importantly, they are do not damage the environment. In the past few decades, thermoelectric generators (TEGs) have emerged as a technology that has garnered substantial attention.

The original version of the chapter has been revised. A correction to this chapter can be found at https://doi.org/10.1007/978-981-95-3921-5_7

A. B. O. Aboukhris
Department of Electrical and Electronics, Near East University, Nicosia, Türkiye

M. Karimzadeh Kolamroudi (✉)
Department of Mechanical Engineering, Near East University, Nicosia, Türkiye
e-mail: Mohammad.karimzadeh@neu.edu.tr

Research Center for Science, Technology and Engineering (BILTEM), Near East University, Nicosia, Türkiye

A. Azizi (ed.), *Smart Diagnostics and Predictive Maintenance*, Emerging Trends in Mechatronics, https://doi.org/10.1007/978-981-95-3921-5_1

TEGs can generate electricity by harnessing waste heat, making them a promising solution for energy recovery and utilization [1]. Furthermore, concentrated solar power systems employ mirrors or lenses, in conjunction with solar tracking systems, to concentrate a vast area of solar radiation and focus it onto a significantly smaller photovoltaic area [2]. Wind, hydroelectric power, geothermal energy, and photovoltaic solar power are also options. Photovoltaic solar energy has experienced remarkable development in the global electricity sector. Expanding policy support is fueling this extraordinary growth, and in 2023 and 2024, China is expected to contribute more to the addition of renewable capacity globally, solidifying its leadership in deployment [3].

Photovoltaic electricity, acquired by converting solar radiation into electricity using photovoltaic cells, can either be fed into the public grid or utilized to power applications in remote locations. While highly accessible and well-suited to resolving the energy crisis, photovoltaic energy is dependent on solar radiation which fluctuates with seasonal changes.

Energy, especially renewable energy sources, holds paramount importance for several compelling reasons:

- Environmental Sustainability: To effectively address climate change and protect the environment, power sources like solar, wind, hydro, and geothermal power offer a manageable substitute for petroleum products.
- Energy Security and Independence: Diversifying the energy mix through the adoption of renewable energy sources reduces reliance on finite fossil fuel resources.
- Economic Development and Employment Creation: The new energy sources acts as a catalyst for economic growth, attracting investments, and generating employment opportunities across various industries, thus stimulating local economies.
- Energy Access and Equity: Harnessing renewable energy sources enables the provision of affordable and clean energy to underserved communities, bridging the energy access gap, and promoting social equity.
- Research and development in areas like energy efficiency, energy storage, and grid integration are all powered by the chase of renewable energy, resulting in overall technological progress, technological advancements and innovation.
- Health and wellbeing renewable energy reduces harmful emissions and air pollution associated with traditional energy sources, contributing to improved air quality, public health, and enhanced community well-being.

In conclusion, protecting the environment, ensuring energy security, ensuring economic prosperity, ensuring equitable access to energy, advancing technology, improving public health, and maintaining cost-effectiveness over the long term all require prioritizing renewable energy sources.

1.2 Motivation of the Study

The motivation behind this study stems from the necessity to optimize grid-connected hybrid PV-battery-supercapacitor systems by employing machine learning techniques for intelligent control strategies. The primary motivation for this research is to develop an intelligent maximum power point tracking (MPPT) technique that incorporates an energy storage system comprising lithium-ion batteries and supercapacitors [4–6]. The secondary motivation for this research includes the integration of PV-based renewable energy sources, the complexities associated with managing energy in such systems, advancements in machine learning, the objective of achieving optimization and improved performance, and the significance of grid stability and resilience. The research aims to develop intelligent control protocols that successfully expand the use of renewable energy, improve framework implementation, reduce costs, and support matrix steady quality using artificial intelligence (AI). This study's findings can possibly contribute to an incredibly energy-productive and green ecosystem future by introducing the development of improved energy storage solutions.

- Machine learning methods can potentially boost control systems' efficiency and performance by employing adaptive algorithms and data-driven models. It is possible to streamline control techniques, and adjust to moving working circumstances by integrating AI into control frameworks. The ease of use and suitability of PID controllers have made them popular in a wide range of applications. Intricate and nonlinear structures may not necessarily give ideal control in all cases. By continuously changing control boundaries and utilizing data, AI algorithms can be integrated with PID controllers to further enhance control solidity and precision [7].
- Integrating hybrid systems and electrical system is more crucial than combining AI strategies. Hybrid systems combine conventional energy sources, such as lattice power, close to energy limit structures, with alternative energy sources, such as daylight-based power sources that are harmless to the ecosystem.

1.3 Problem Statement

Solar energy is an important, long-term source of energy that has a great deal of potential to address global power supply issues. Using less fossil fuel provides a practical method for lowering greenhouse gas production and contributing to the planet's long-term sustainability. However, a significant obstacle is the intermittent nature of solar energy, which is influenced by factors like the amount of cloud cover and the time of day. The intermittent nature of solar energy prevents its widespread use. The sunlight-based power source is diminished or totally halted in the evenings or cloudy weathers. This variation in energy production may cause the power grid to become unstable, necessitating additional power sources. To maximize solar energy output, watching out for this irregularity through intelligent fixes is essential. The output of a photovoltaic array is predominantly influenced by solar irradiation and solar cell temperature, resulting in varying power levels. However, the current electric power grid cannot accommodate such fluctuations in the power supply from renewable sources. Thus, the irregular nature of solar power poses a tremendous integration challenge when directly connected to the utility network. To address this issue, there is a pressing need to develop low-cost, high-capacity storage systems. These structures can deliver solar power at a reliable level over an extended time span for utility applications. By bridging the gap between the demands of the utility grid and solar energy generation, such storage systems would ensure a consistent and uninterrupted supply of power.

Managing the interaction between supercapacitors and batteries in an energy storage system can be challenging, particularly with respect to fluctuations in battery charge and discharge. These fluctuations can impact the lifetime and performance of the battery over the long term. One of the primary concerns is the possibility of high current spikes during rapid charge or discharge processes. The stress from these current spikes can strain the battery cells, leading to increased heating, altered voltage characteristics, and accelerated degradation. Reduced battery capacity, decreased efficiency, and overall shortened lifespan can result from repeated exposure to such fluctuations. When a supercapacitor is connected to a battery, it can quickly store or deliver high-power bursts due to its low internal resistance.

1.4 Solution of the Problems

Combining these two components in an energy storage system is highly advantageous because it leverages their distinct characteristics: the high-power density of supercapacitors and the high-energy density of batteries. Supercapacitors excel at rapid charge and discharge rates, albeit with lower energy density, whereas batteries perform well at slower charge and discharge rates. Nevertheless, exposing the battery to fast-varying (high-frequency) power components during its charging or discharging cycles can shorten its lifespan. By incorporating a supercapacitor into the storage system in close proximity to the battery, it is possible to divert these rapidly changing power components away from the battery bank, thereby extending its lifetime. Furthermore, the battery's lifespan is profoundly affected by discharging it beyond its Depth of Discharge (DoD) limit. To extend the battery's expected lifespan, it is advisable to avoid discharging it beyond this DoD threshold. The key idea underlying this arrangement is to exploit the strengths of each component: the battery provides support for low-frequency power components, while the supercapacitor handles high-frequency power components through conventional controllers. However, existing control strategies for such hybrid energy storage configurations have limitations. They may be vulnerable to worst-case scenarios, involve complex control algorithms, and require extensive numerical computations. The objective of this approach is to improve the control algorithm, reduce computational complexity, and enhance the overall robustness of the system.

1.5 Aim of the Study

Using AI techniques, the goal of this investigation is to optimize the control strategy for grid-connected hybrid PV-battery-supercapacitor systems. The investigation is structured around three primary objectives: developing an intelligent MPPT methodology for maximum power point tracking, applying artificial intelligence algorithms for energy management to optimize energy flow and utilization, and employing direct-quadrature control for seamless integration of the hybrid system with the grid, thereby ensuring stable power transfer and grid robustness. By achieving these objectives, the study aims to enhance the performance, energy efficiency, and grid integration capabilities of hybrid PV-battery-supercapacitor systems. It will also contribute to the development of intelligent control strategies for renewable energy systems and support the transition toward a more sustainable energy future. Ultimately, the goals are to devise smart control strategies that improve grid stability, system performance, and energy efficiency.

1.6 Research Objectives

Employing AI techniques to achieve effective control and maximization of the system, enhance energy efficiency, and further improve reliability and stability are among the goals. The study focuses on designing and simulating novel control algorithms that balance power flow, optimize energy management, and seamlessly integrate with the grid for a hybrid system composed of PV, lithium-ion battery, and supercapacitor. In addition, the research aims to enable intelligent decision-making and adaptive control through machine learning. The proposed control strategies will be validated through simulations conducted in MATLAB/SIMULINK, distinguishing them from existing methods. The outcomes of the study will contribute to the advancement toward a future powered by renewable energy and promote smart control strategies for its integration.

1.7 Research Outline

This research is organized as follows: Sect. 2 provides a comprehensive literature review that examines existing research on related topics to identify optimal energy management and control techniques for grid-connected battery-supercapacitor systems. Section 3 presents the methodology, detailing various control strategies employed to facilitate effective energy management for the hybrid system's integration with the electrical grid. Section 4 focuses on validating the efficacy of the proposed methods and control systems through rigorous simulations and testing. Finally, Sect. 5 summarizes the findings and discusses potential directions for future research.

2 Literature Review

2.1 *Introduction*

Globally, there is an increasing recognition of the imperative to transition from fossil fuels to environmentally sustainable energy sources. This shift is driven by environmental concerns, energy security, and economic considerations. Sustainable energy encompasses a variety of technologies and resources, including solar, wind, hydroelectric, geothermal, and biomass energy [8–10]. This introduction provides an analysis of the global progression of renewable energy, highlighting key milestones, challenges, and the potential for a sustainable energy future. Solar energy, in particular, offers numerous advantages, whether utilized in off-grid systems [11–14]. Due to its low cost, wide availability, and ease of production, solar energy is extensively utilized. Solar panels provide an efficient method for generating and converting sunlight into electrical energy [15]. Photovoltaic systems are particularly beneficial in remote regions where grid connection is impractical, providing energy autonomy in decentralized configurations. Off-grid systems facilitate access to power in remote areas, leading to ongoing reductions in energy costs and delivering long-term environmental benefits through the supply of clean, sustainable energy. Additionally, these systems enhance reliability and resilience by ensuring a consistent power supply during outages or natural disasters. Grid-connected solar systems also offer further advantages [16]. Net metering and feed-in tariffs, which allow excess power to be fed back into the grid for financial compensation, are enabled by solar systems integrated with the grid.

During periods of heightened demand or diminished sunlight, these systems draw from the grid to ensure a stable supply. Additionally, users of grid-connected solar systems offset their consumption against the grid, potentially accruing surplus energy credits, resulting in reduced utility bills. Grid-connected structures are both flexible and adaptable, thereby facilitating their ability to accommodate varying energy requirements. They also promote the integration of renewable energy sources into the energy matrix, thereby contributing to an environmentally sustainable future [17]. In summary, grid-connected and standalone solar energy systems present distinct benefits tailored to their respective contexts and energy needs. In remote areas, independent systems provide autonomy and reliability, whereas grid-connected systems offer resilience and economic benefits while facilitating progress toward a more sustainable energy landscape [18–22]. The criteria of power and energy needs, response time, lifespan, and cost-effectiveness all play a role in determining the optimal energy storage system. Due to their high energy density and adaptability, batteries are frequently employed in both commercial and private settings. In contrast to batteries, supercapacitors can deliver rapid bursts of power and possess longer lifespans. Flywheel energy storage systems employ the principle of kinetic energy accumulation and are recognized for their rapid response times and high cycling capabilities. The overall strength and reliability of the energy system are enhanced when these energy storage innovations are integrated with solar panels. By storing excess energy and releasing it when demand is high or sunlight is scarce, it exploits solar energy to its fullest potential [23]. This approach not only ensures a more stable power supply but also reduces reliance on traditional methods of energy generation that use non-renewable resources. Integrating energy storage systems with solar power will be instrumental in realizing a cost-effective energy future as the demand for sustainable power sources continues to increase. By effectively harnessing and utilizing solar energy, coupled with energy storage, we can prepare for a more sustainable and robust energy grid [24–26].

2.2 Generation Capacity of Photovoltaic Energy Sources: Strategies and Advancements

The generation capacity of renewable energy sources refers to the maximum potential electrical output derived from the utilization of resources that are naturally replenished, such as solar irradiance, wind currents, hydrological cycles, and geothermal activity. The capacity of an eco-friendly energy system is a critical indicator of its potential to deliver power to the grid [27]. Given their capacity for natural restoration, renewable energy sources represent environmentally viable alternatives to conventional fossil fuels. Modifications to capacity limits take into account factors such as geographical location, climatic conditions, and technological advancements. Due to the inherent variability and intermittency associated with renewable energy sources, along with the necessity for an expanded grid infrastructure and advanced energy storage technologies, the advancement of sustainable energy faces numerous challenges [28]. However, ongoing research, innovation, and collaboration among governments, corporations, and academia are actively addressing these challenges. These endeavors establish a foundation for a more sustainable energy future [25], as illustrated in Fig. 1 [29].

The development of renewable energy holds enormous potential. It offers the opportunity to mitigate the adverse effects of climate change by significantly reducing greenhouse gas emissions, enhancing energy security, and promoting economic growth [30]. To realize this potential and successfully transition to a fully sustainable energy economy, it is essential to provide continuous support, undertake major projects, and foster collaboration among various stakeholders. Furthermore, ensuring policy stability and consistency for all renewable energy and efficiency technologies

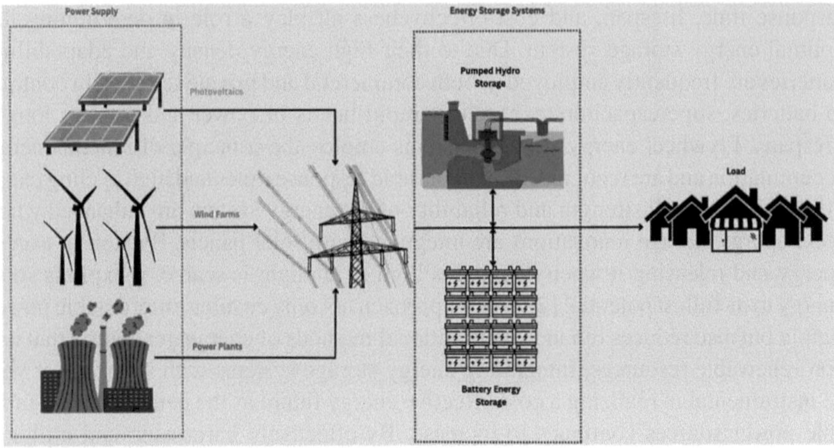

Fig. 1 Renewable energy for sustainable growth [29]

is crucial to fully exploit the benefits of sustainable power and to create a future energy framework that is more affordable and adaptable [31].

2.3 *Unleashing the Potential of Photovoltaic Energy Sources: Accelerating Growth in Power Capacity*

There has been a significant transition toward clean and sustainable energy sources, as demonstrated by the substantial global adoption of renewable power capacity. This growth is evidenced by the extensive application of solar and wind power technologies across various nations and regions worldwide [32]. The increasing recognition of the ecological and economic benefits of sustainable energy is apparent in the rise of solar and wind power infrastructure installations. Solar power exploits the abundant energy from the sun, while wind power harnesses the kinetic energy of the wind, offering pristine and inexhaustible alternatives to conventional fossil fuel-based power generation. This global shift toward sustainable technologies is driven by several factors. Firstly, solar panels and wind turbines have become more economically viable and competitive with traditional energy sources due to decreasing costs [33]. Additionally, government policies and incentives, such as feed-in tariffs and renewable portfolio standards, have played a vital role in facilitating the widespread adoption of renewable energy. With the anticipated initiation of new solar and wind projects in 2023, these two energy sources are projected to account for 16% of the total generation, marking an increase from 14% in the previous year and 8% in 2018. Conversely, our projections indicate that the share of generation from coal will decline from 20% in 2022 to 18% in 2023, while the share from natural gas will decrease from 39 to 38%. The swift development of renewable energy resources,

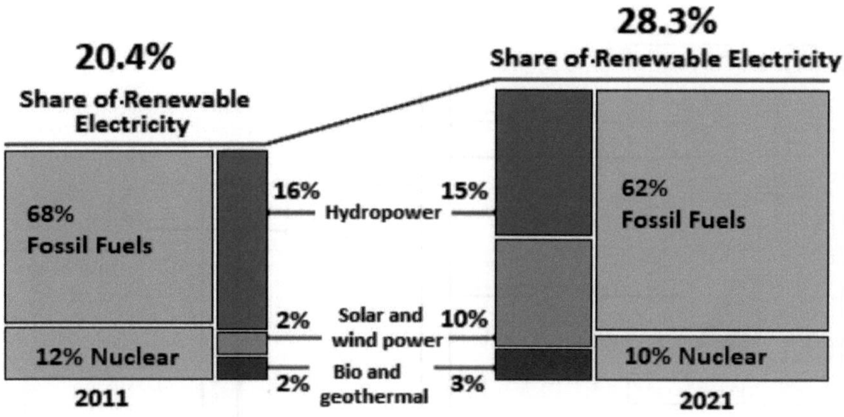

Fig. 2 Electricity consumption by energy source in the years 2018–2021 [34]

particularly in the domains of solar and wind, has been one of the most significant alterations to the composition of energy sources used for electricity generation in recent years, as reflected in Fig. 2 [34, 35].

2.4 Solar Photovoltaic Systems with Storage Sources

Integrating energy storage with solar power substantially enhances system reliability and effectiveness, making solar energy systems more adaptable to varying energy demands and grid conditions. Currently, battery-dependent storage systems show significant promise in mitigating demand-generation oscillations in large-scale remote area power supply applications (RAPSAs). Lead-acid and lithium-ion batteries are the most used energy storage solutions in standalone photovoltaic systems [36]. These battery-designed storage systems exhibit high energy density but relatively low power density, resulting in limited charge/discharge rates [37]. In terms of performance, lead–acid batteries exhibit considerably higher loss rates, ranging from 15 to 20%; on the other hand, lithium-ion batteries typically experience lower losses, with rates varying from 2 to 8%. In comparison with lithium-ion batteries, lead-acid batteries have longer charged times. Inadequate and irregular charging, particularly during winter, can significantly shorten the lifespan of lead-acid batteries. In contrast, supercapacitors (SCs) boast high-power density but lower energy density, typically leading to more frequent charge/discharge cycles. Existing literature offers a comprehensive performance comparison between batteries and SCs [38, 39]. This combination allows the system to store excess energy generated during periods of sunlight for later use, providing a more reliable and continuous power supply. The studied system components include solar panels, inverters, energy

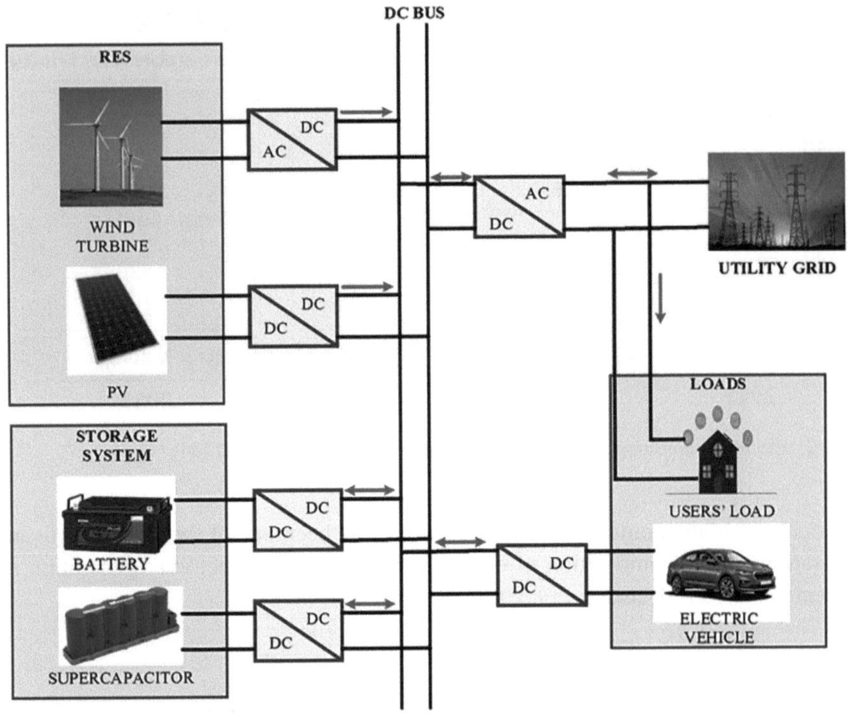

Fig. 3 Typical hybrid energy storage-electrical grid design [41]

storage systems, charge controllers, and a power management system as addressed in Fig. 3 [40, 41].

It is important to emphasize that the latter system incorporates power converters, notably the boost converter. In PV systems in particular, the boost converter is essential to the integration of renewable energy sources. It enables the extraction and conversion of the maximum power available from the PV panels, allowing for efficient utilization of the generated energy. By regulating the voltage and current levels, the boost converter ensures that the load receives the required power while minimizing losses as addressed in Fig. 4.

Another critical component is the MPPT controller, such as perturb and observe, incremental conductance, fuzzy logic, artificial intelligence, and particle swarm optimization, as presented in Figs. 5, 6, 7, 8, and 9. According to Mokhlis et al., the MPPT controller is essential for maximizing the power output of the renewable energy source, especially in PV systems [4].

A Buck-Boost converter is used to charge and release the stockpiling source. This converter allows power to flow in either direction between the storage source and the distribution network, as depicted in Fig. 10.

The integration of the hybrid system with the electrical grid is achieved by employing a Phase-Locked Loop (PLL) (Fig. 12 [50]) in conjunction with a DC/

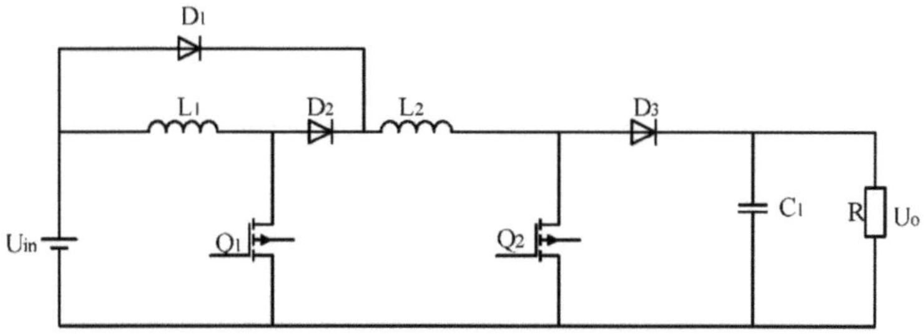

Fig. 4 Boost converter topology [42]

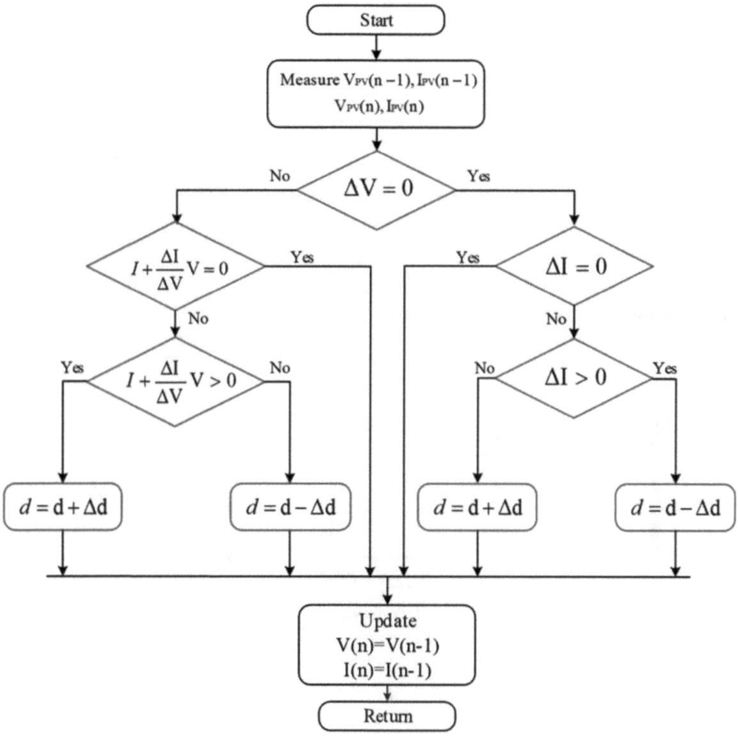

Fig. 5 Incremental conductance method [43]

AC converter (Fig. 11 [49]). These components are pivotal in transforming the direct current (DC) output of the hybrid system into alternating current (AC) suitable for grid use. The DC power generated by the hybrid system must be converted into AC power by means of a DC/AC converter, occasionally referred to as an inverter. This conversion is executed utilizing power electronic components such as silicon carbide (SiC) devices or insulated gate bipolar transistors (IGBTs). The converter ensures that the AC power generated by the hybrid system aligns with the grid's

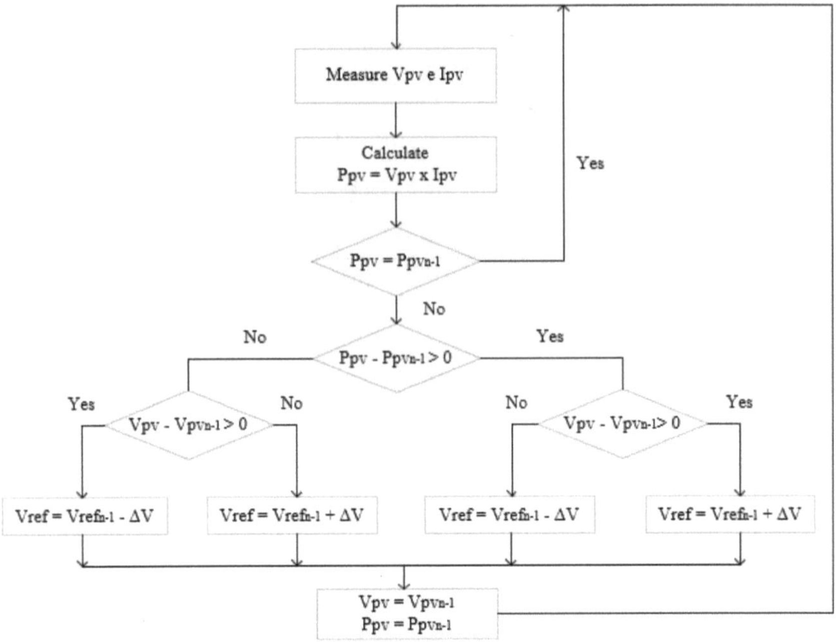

Fig. 6 Perturb and observe method [44]

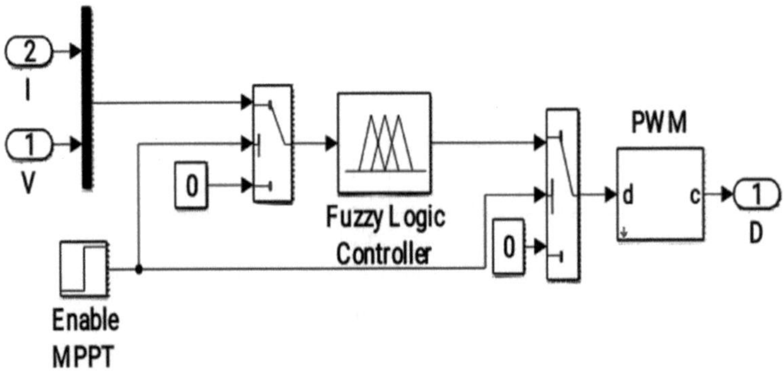

Fig. 7 Fuzzy logic method [45]

voltage, frequency, and phase requirements. The PLL aligns the AC output of the hybrid system with the voltage and frequency of the grid. It continuously monitors the grid's voltage waveform and adjusts the AC output's phase and frequency to harmonize with the grid. This synchronization is critical for the reliable and stable integration of the hybrid system with the grid, ensuring that the integrated system operates in unison with the existing grid infrastructure. Through the utilization of a

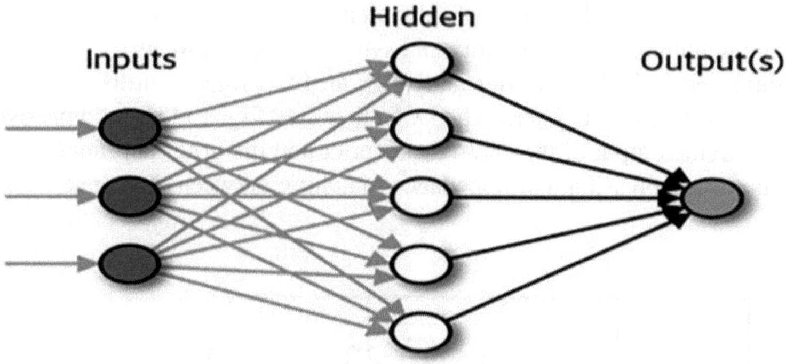

Fig. 8 Machine learning algorithm [46]

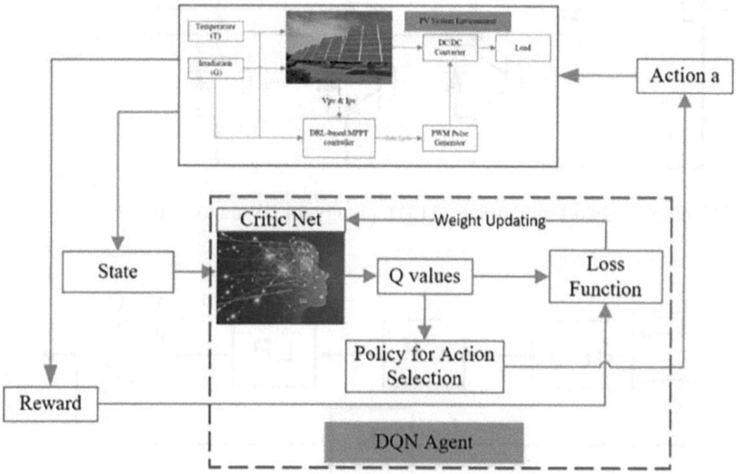

Fig. 9 Deep learning algorithm [47]

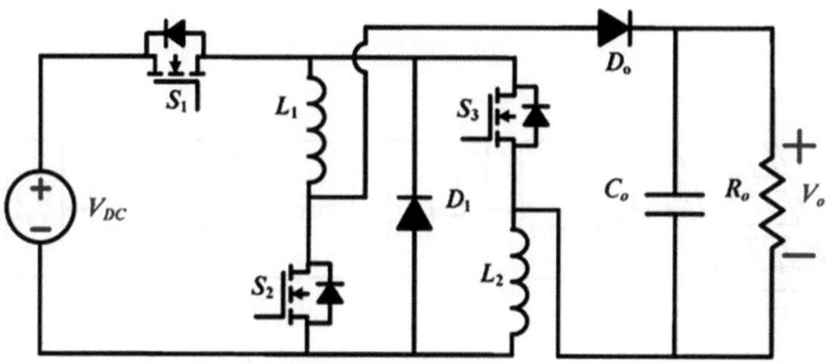

Fig. 10 Buck-Boost converter [48]

DC/AC converter and a PLL, the hybrid system can efficiently transmit power to the grid while maintaining compatibility and synchronization [40, 49].

Through the DC/AC converter and PLL, an LCL filter is utilized for the grid integration of the hybrid system. The LCL filter (see Fig. 13 [51]) is a passive filter connected to the output of the DC/AC converter and the grid. Its primary functions are harmonic suppression and impedance matching between the converter and the grid.

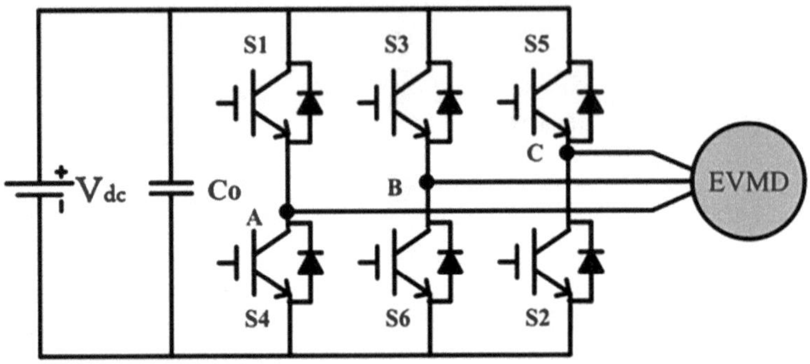

Fig. 11 Three-phase two-level inverter (TLI) topology for EV motor drive [49]

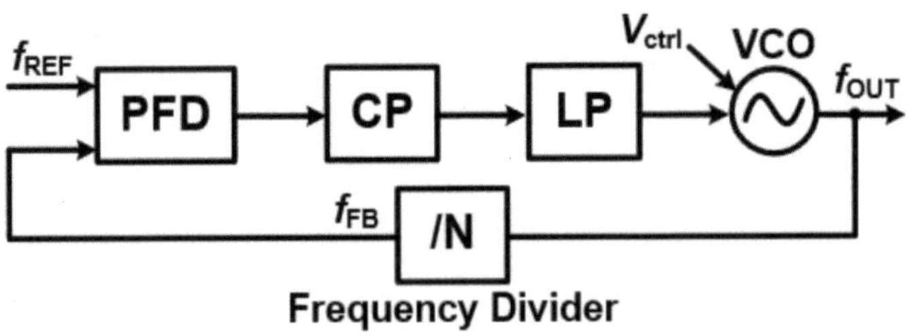

Fig. 12 PLL model [50]

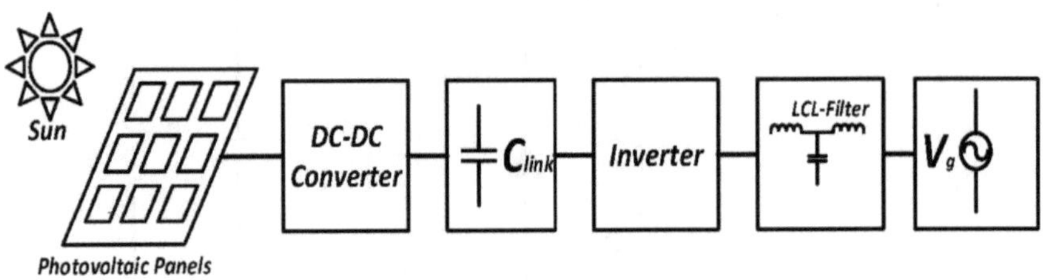

Fig. 13 LCL filter [51]

2.5 Previous Studies and Analysis of PV and Storage Sources

Mechanical advancements and an evolving energy landscape create promising opportunities for photovoltaic (PV) storage systems. These systems offer numerous benefits, such as increased energy independence, grid support, and reliability. The ability to store excess energy during periods of high production provides a dependable power source during grid outages, enhancing resilience. Another advantage of PV storage is peak load shaving, which reduces costs and energy consumption during high-demand periods. Smart grids integrated with photovoltaic (PV) storage improve energy management and demand response efficiency [52]. Advancements in battery technology, such as lower costs and higher energy density, further enhance the performance and financial viability of PV storage systems. The decentralization of energy production empowers individual users to generate, store, and manage their power, contributing to a more robust and adaptable energy landscape. Additionally, government policies and incentives drive the widespread adoption of PV storage, fostering a supportive regulatory environment. As costs continue to decline and research and development advance, the integration of photovoltaic systems with energy storage holds significant potential for transforming the energy sector, promoting sustainability, and enhancing overall efficiency. Within the evolving energy landscape, photovoltaic storage plays a vital role in the transition to a more reliable, sustainable, and environmentally friendly energy future.

2.6 Advances in PV and Storage Technologies

Recent innovations in PV and energy storage technologies are transforming the energy landscape and pushing it toward greater integration, sustainability, and efficiency. In the field of PV, huge progress has been made in upgrading the proficiency and cost viability of solar systems. Further developments in assembling procedures and material science have potential for higher conversion efficiencies, empowering PV systems to produce greater power from the same amount of solar irradiance. This has made solar energy more accessible and economically viable, paving the way for widespread adoption [53].

2.7 Advances in PV Technology

When light photons are exposed to an electrical device, such as a PV cell, electricity is produced. French physicist Antoine Becquerel discovered the PV effect in 1839. Although the technology did not see widespread adoption until the 1960s. It was not until the late 1980s that solar panels, composed of cell assemblies, began appearing on rooftops. The growth of PV technology, marked by the development of large-scale power generation parks, has continued unabated since the early twenty-first century [54].

(A) Operation principle

The operation of a solar cell system relies on semiconductors, which, when exposed to photons, generate an electron flow. Photons are described by Albert Einstein in the 1920s as the "grains of light," carry solar energy at a speed of 300,000 km/s. When photons strike a semiconductor material, such as silicon, they dislodge electrons from its atoms. These freed electrons move chaotically, seeking "holes" to fill [55].

To establish an electric current, electron movements must be aligned in the same direction. This is facilitated by using two types of silicon. One side is doped with phosphorus atoms, which have more electrons than silicon, while the other side is doped with boron atoms, which have fewer electrons. This dual-sided configuration functions like a battery: the side with excess electrons becomes the negative terminal (N), while the side with fewer electrons becomes the positive terminal (P). An electric field forms between these terminals. When photons excite electrons, the electric field directs them toward the N zone, while "holes" migrate toward the P zone [56]. These electron and hole movements are collected by electrical contacts placed on the surfaces of the two zones. As the collected charge flows into an external circuit, a direct current is produced. An anti-reflective layer is applied to minimize photon loss due to surface reflection. Figure 14 illustrates the PV cell operation [57].

(B) Photovoltaic cell efficiency

Efficiency is measured as the electrical power generated on a solar cell divided by the incident light power. To achieve this, the cells are meticulously calibrated through a process involving their assembly into modules and subsequent incorporation into panels. Calibration is done by placing these panels in front of a solar simulator, which replicates optimal conditions: 1000 W of light per square meter of sunlight and a temperature of 25 °C [58].

Under real-world conditions, the actual power output, referred to as "producible," is determined by considering various factors, including the cell's output, the average sunlight levels in the region throughout the year, and the specific conditions of the installation.

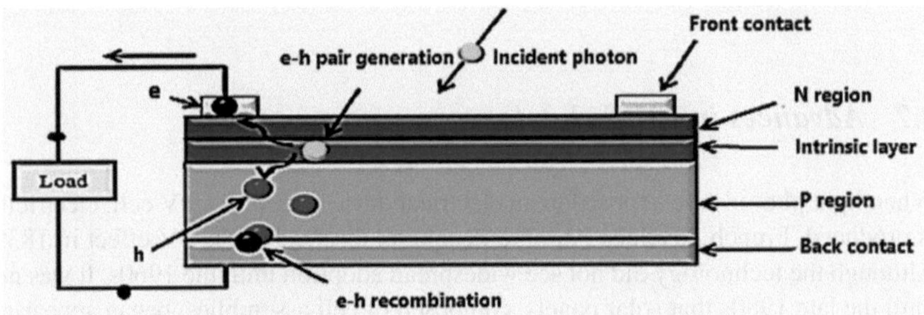

Fig. 14 PV cell operation [57]

(C) **Photovoltaic cells types**

- ***Crystalline silicon***

Silicon, a key material in solar cell production, is extracted from silica, with quartz being one of its abundant forms commonly found in sands. Silicon cells (see Fig. 15) dominate the market, constituting over 90 percent of the industry. Despite significant advancements in PV systems, the conversion efficiency of solar cells remains relatively low. For example, silicon solar cells and multi-junction types have maximum power conversion efficiencies of approximately 20% and 30%, respectively, indicating further room for improvement in enhancing their ability to convert solar radiation into electrical power [59]. The average efficiency of commercial silicon cell products varies between 14 and 22%, depending on the technology employed [60].

Through cold treatment, silicon undergoes a process leading to the formation of multiple crystals (polycrystalline). This method, known for its simplicity, boasts a laboratory yield surpassing 22%. Alternatively, molten silicon can be transformed into a single large crystal (monocrystalline) with an efficiency reaching up to 26.6%. Notably, the cost competitiveness of these silicon cells with other energy production solutions has significantly increased in recent years [61].

- ***Thin film PV cells***

Instead of the traditional method of cutting silicon into thin slices of about 200 microns, there is an alternative approach where semiconductor materials are deposited in layers, each a few microns thick, onto a substrate such as glass or plastic. Materials like cadmium telluride or CIGS (copper/indium/gallium/selenium) can be

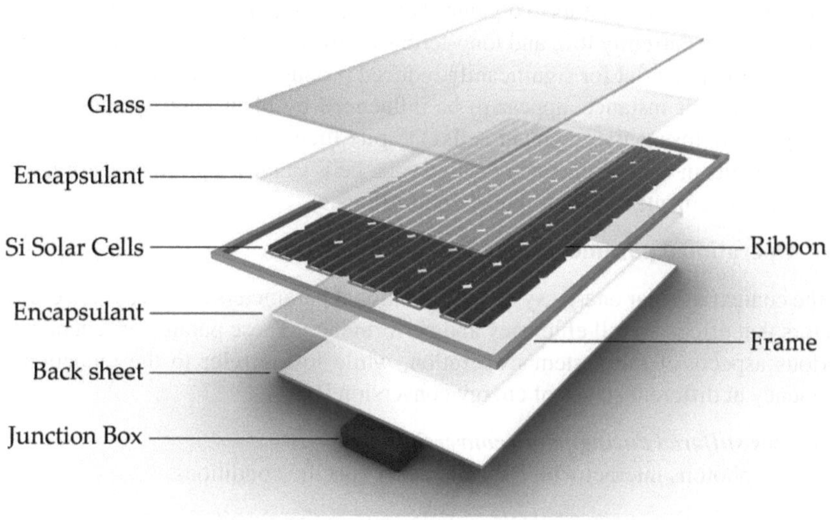

Fig. 15 Crystalline silicon PV cell [62]

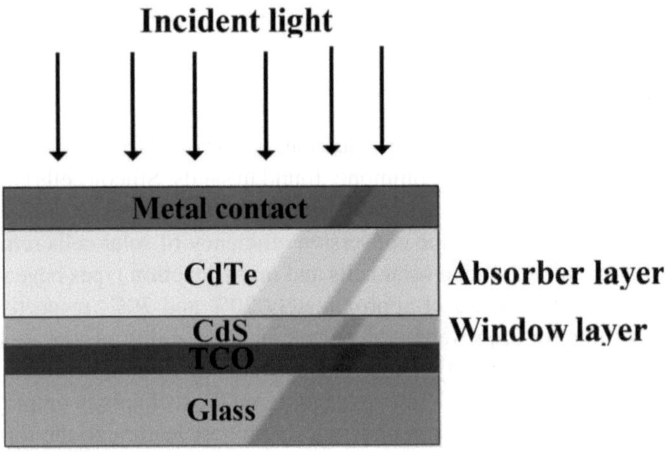

Fig. 16 Cross section of thin films PV cell [64]

utilized for this purpose. Laboratory yields for these materials are comparable to silicon, with rates of 22.1% and 23.3%, respectively [63]. Thin films cell can be seen in Fig. 16 [64].

Another option involves using silicon in its "amorphous" (uncrystallized) form for thin films. While this technology has long been used in small calculators, it performs less well than more conventional silicon-based techniques.

- ***Organic PV cells***

Organic PV cells, shown in Fig. 17 [65], are gaining traction as they shift their focus from mineral semiconductors to organic chemistry molecules or polymers. Although their yields are currently low, and long-term stabilization remains a challenge, these cells hold the potential for significantly reduced production costs [61].

Dye cells, for instance, appear to be influenced by plant photosynthesis, functioning as photosensitive pigment cells. Despite their current limitations, ongoing advancements in organic chemistry-based solar cells suggest a promising avenue for future improvements and widespread applications.

(D) **Operational parameters and losses**

In the context of solar energy systems, operational parameters and losses are crucial factors that affect overall efficiency and performance. These parameters encompass various aspects of the system's operation, while losses refer to the reduction in efficiency at different stages of energy conversion [66].

- ***Losses suffered during photo conversion***

  When photons interact with PV cells under specific conditions:

  - They generate pairs of electron–hole pairs in the circuit.

  This process contributes to:

Fig. 17 Organic PV cells on a greenhouse roof [65]

- Enhancing the efficiency of the PV panel.

Fundamentally, when photons interact with photovoltaic cells, they instigate the formation of electron–hole pairs, a phenomenon that ultimately enhances the overall efficiency of the PV panel. However, in practical terms, numerous factors can influence the efficiency of photo conversion [67].

One such factor is spectral loss, where the wavelength of photons plays a critical role. The wavelength must be sufficiently short to ensure that the photon's energy exceeds the material gap and is effectively absorbed. For instance, an ultraviolet wavelength photon carries more energy than a visible radiation photon, and gamma rays exhibit even higher energy levels. To illustrate, considering a silicon gap of 1.1 eV, the sample absorbs only photons with wavelengths less than 1100 nm, aligning with the optimal absorption zone dictated by the gap's energy [68]. The second influencing factor is that not all photo-generated carriers are successfully recovered. When electrons and holes separate in the depletion zone, they disperse into the material's neutral zones. Due to recombination, some of these charges fail to contribute to the creation of current, thus hindering efficient energy conversion. It is crucial to emphasize that the profitability of PV conversion depends on factors such as leakage current, condition losses, and radiation energy. Addressing these considerations is pivotal for optimizing the efficiency and effectiveness of photovoltaic systems.

2.8 Advances in Storage Technology

Advancements in energy storage technology have improved the efficiency and reliability of PV systems. The integration of advanced storage solutions addresses the intermittent nature of solar power generation, enabling the capture and utilization of excess energy during periods of high sunlight. This stored energy can be used in low- or high-demand situations to ensure a consistent and reliable power supply [69]. Technological innovations, such as advanced battery systems and smart grid technologies, are critical to increasing the energy storage capacity of PV systems. High-capacity batteries with improved energy density and longer life cycles enhance storage capabilities, allowing for better management of fluctuations in energy production and consumption [70]. Additionally, advancements in system integration technologies enable seamless communication between PV systems and the broader electrical grid. This integration facilitates efficient utilization of solar energy and strengthens grid resilience. The development of sophisticated energy management systems further improves the control and coordination of energy flows, enabling users to monitor, store, and distribute solar energy with greater precision.

2.9 PV Solar Energy Integration into the Electrical Grid or at a Remote Location

The incorporation of solar PV energy into electrical systems holds significant economic and energy importance. The method of integration primarily depends on the characteristics of the electrical system in question, such as whether it operates independently (autonomous), is linked to the electrical distribution grid, or incorporates a supplementary renewable source. Consequently, there are three distinct approaches to integrating photovoltaic solar energy: autonomous, grid-connected, and hybrid systems [4].

2.10 Grid-Connected Solar Systems

Grid-connected PV systems are, above all, production equipment designed to generate photovoltaic electrical energy that will be injected into a power grid supplied by other generation sources, such as high-capacity rotating machines. Depending on whether the energy produced is utilized on the production site or not, two electrical connection topologies can be explored [71].

- ***The entire injection without any storage***

Solar radiation undergoes direct conversion into energy through a photovoltaic generator. Subsequently, a PV inverter transforms the generated direct current from solar modules into alternating current, aligning it with the electrical parameters of the public distribution grid. The entirety of the produced power is seamlessly injected into the public distribution grid, enabling the producer to receive compensation under optimal economic conditions. In the event of an electrical grid failure, characterized by voltage loss or frequency deviation beyond tolerance limits, the PV inverter automatically ceases operation.

- ***Total injection including storage***

The fundamental concept persists, yet the manufacturing process now encompasses energy storage using accumulators. However, due to economic considerations, the storage capacity is restricted and aimed at facilitating smoother production and/or replenishing energy reserves by day's end to manage peak electricity grid demands.

- ***Self-consumption without storage***

The photovoltaic system maintains its regular operation and is directly connected to the electrical infrastructure of the facility. The receivers operate autonomously, consuming solar electricity generated on-site. Any surplus solar energy is fed into the public distribution system. In the event of an electrical grid failure, characterized by a loss of voltage or a frequency deviation beyond tolerance limits, the photovoltaic inverter ceases operation, halting the supply of electricity to the receivers [72–74].

- ***Self-consumption including storage***

The concept remains consistent with the previous description, but now the photovoltaic system incorporates accumulators, enabling the storage of energy generated during the day for subsequent local consumption in the evening. However, due to cost considerations, the storage capacity is limited to a few hours of operation and it is not completely independent from the electrical grid. This approach offers the same advantages as the previous one but with more dependence on solar power. Furthermore, this configuration autonomously switches to solar power for a brief periods of time, acting as a backup in the case of a grid failure. However, it is important to note that this system has the drawback of electrochemical storage, leading to associated costs for investment, maintenance, renewal, and recycling, among other factors.

2.11 *Off-Grid Photovoltaic Systems*

Autonomous systems relying solely on photovoltaic energy can be distinguished from hybrid systems within the domain of off-grid photovoltaic systems. Hybrid systems integrate one or more supplementary energy sources, such as renewable sources like wind or hydropower, or non-renewable generators. The primary objective of these off-grid systems is to fully meet user needs. Thus, the term "photovoltaic system" encompasses all aspects, including production, storage, management, conversion, distribution, and utilization. These systems are divided into two categories: those operating without energy storage and those using accumulators to store energy [75].

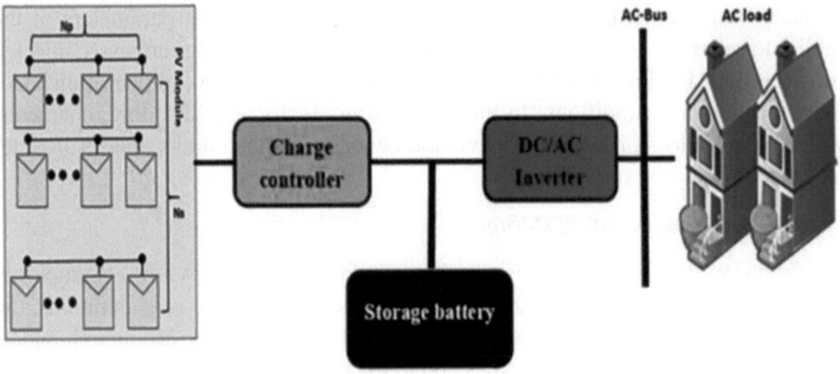

Fig. 18 Standalone PV system with storage [77]

- ***PV systems without storage***

Throughout the day, the solar generator supplies power to users either directly or through a DC/DC or DC/AC converter, depending on available sunlight. The amount of power allocated to users depends on prevailing sunshine levels. At night, the system is deactivated as it no longer receives sunlight. Various photovoltaic applications, such as pumping, ventilation, refrigeration, and seawater desalination, are among the most common uses of solar energy [76].

- ***Autonomous PV systems with storage***

The operational principle is as follows: During daylight hours, the PV generator supplies power to the direct current receivers and simultaneously charges the storage battery, as illustrated in Fig. 18 [77]. A charge controller is in place to protect the battery and prevent overcharging in solar overproduction situations. During nighttime or adverse weather conditions, the receivers draw power from the batteries. In the event of a deep discharge, a discharge limiter acts as a safeguard for the battery. An independent inverter transforms direct voltage into alternating voltage when AC receivers are present, enabling their power supply.

2.12 Hybrid PV Systems

Depending on the inclusion of energy storage, two electrical topologies can be considered for hybrid systems:

- Photovoltaic installations with accumulators, coupled with a renewable source such as wind power or micro hydraulics.
- Photovoltaic systems with generators that do not incorporate energy storage.

- ***Hybrid PV system with storage***

Throughout the day, the solar generator initiates the charging process while simultaneously supplying power to the inverter. The inverter, in turn, converts direct voltage to alternating voltage, delivering electricity to the alternating-mode receivers [78]. To prevent overcharging during periods of solar overproduction, a charge controller safeguards the battery. In instances of insufficient sunlight or excessive receiver usage, a backup generator steps in to provide direct power to the receivers while also replenishing the battery, ensuring uninterrupted service, as illustrated in Fig. 19 [79].

- ***Hybrid PV system without storage***

The generators continuously provide electrical power to the distribution grid. Depending on the sunlight availability, the PV generator injects varying amounts of electricity into the grid, as depicted in [75, 76]. The PV generator is not obligated to supply the entirety of the energy it produces, resulting in reduced fuel consumption and operational costs. Hybrid systems with storage, ranging from a few kW to a few hundred kW, prove beneficial for individual residences, mountain chalets, high-power telecommunications relays, and community electricity needs. On the other hand, hybrid systems without storage, ranging from a few tens of kW to a few MW, are primarily employed for electrifying large settlements and powering mines engaged in extracting high-power raw materials [80, 81]. The implementation of an

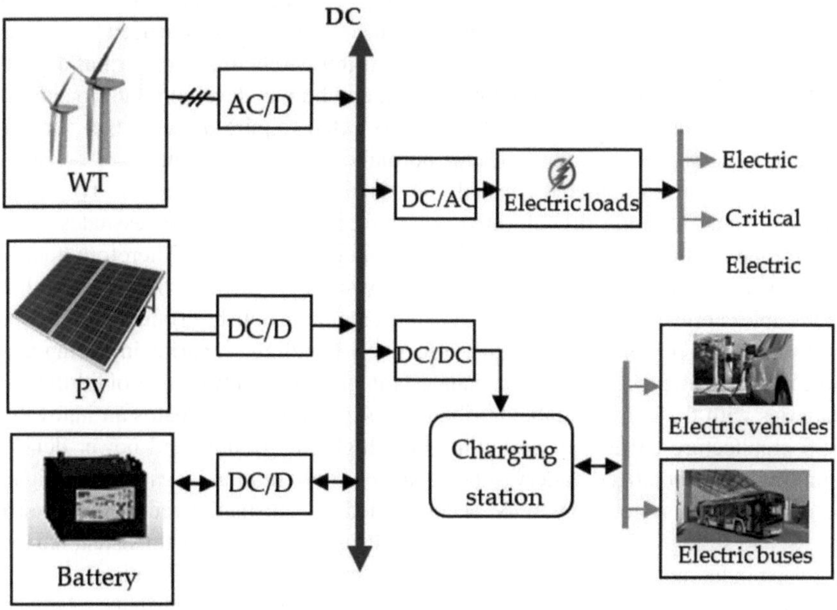

Fig. 19 Hybrid PV system with storage [79]

energy management control system is imperative to ensure the continuity of the PV system and extend the lifespan of the energy storage system.

By acting as a central control mechanism, an energy management control system plays a crucial role in optimizing the operation and coordination of various components within the system. It enables efficient energy flow management, resource allocation, and performance optimization.

2.13 Improving Energy Management Strategies

This work presents a Hybrid Energy Storage System that integrates supercapacitors and batteries into the framework. The effective coordination of energy flows among these components plays a critical role in enhancing the overall performance of the system. Accordingly, our research focuses on developing an intelligent energy management strategy tailored to optimize the functionality of the proposed hybrid system. To provide a comprehensive understanding of energy management, this section includes a literature review of existing energy management strategies. Various approaches and methodologies related to energy management in hybrid systems have been explored. The proposed energy management strategy incorporates batteries and supercapacitors specifically designed for the hybrid system. According to Hredzak, [82], the objectives of this strategy are to ensure reliable operation, increase system efficiency, and maximize the utilization of energy resources. Today's pressing issues, such as energy demand and climate change concerns, underscore the need for sustainable and balanced energy solutions, as energy is intricately linked to various aspects, including the environment, economy, society, and human development. Loads such as motors, air conditioners, and refrigerators often require a significant surge of current, typically seven to ten times the normal operational current.

This high current demand can only be sustained for a short period, typically a few moments. Meeting this elevated power requirement often necessitates costly and frequent battery resizing, typically every three to five years. Therefore, there is a compelling case for using an optimal energy storage system (ESS) in standalone renewable energy systems. Such a system must maintain both high energy throughput and substantial power capacity while effectively handling scenarios involving variable loads and unexpected weather changes. Integrating the advantages of lithium-ion batteries with the commendable attributes of supercapacitors creates an innovative ESS architecture capable of sustaining both high energy flow and power density. Employing a specialized battery-supercapacitor configuration is expected to significantly enhance system longevity, accompanied by substantial reductions in size, stress, and storage costs. This framework is designed to ensure a consistent balance between input and output power (i.e., between production and utilization).

A diverse array of control algorithms for managing power distribution between supercapacitors (SCs) and batteries has been extensively documented [83–88]. These control methods include neural networks, rule-based models, model predictive control, and fuzzy logic [89]. A detailed analysis of how incorporating an SC storage device can extend battery lifespan is provided by [87]. A hybrid remote area power supply application (RAPSA) using wind power explores the benefits of integrating SCs into a battery-based storage system. Consequently, an energy management algorithm (EMA) has been developed to control both the battery and SC storage system in a predefined manner. Specialized applications related to Hybrid Energy Storage Systems (HESS) within microgrids are documented in [90]. Kadri et al. demonstrate how HESS can reduce battery costs while enhancing overall system efficiency [84]. Ortúzar et al. use Artificial Neural Networks (ANNs) to develop and control a HESS for electric vehicles [85]. Additionally, a rule-based control system for integrating battery storage with solar and wind energy sources is detailed by [86].

However, the advanced solutions mentioned have certain limitations, primarily related to the need for extensive data storage capacity and computationally intensive processes. A specific model predictive control (MPC) method suitable for hybrid power sources is presented in [91]. Nevertheless, one of its main limitations is the complex mathematical processing involved due to its strong dependence on traditional MPC. Chong et al. evaluate the technical and financial characteristics of various standalone HESS structures for Renewable Energy Power Systems (REPS), including active, semi-active, and passive HESS [87]. Additionally, researchers [89, 90, 92] have examined a distinct area of passive HESS in relation to the generation of the Pulse Width Modulation (PWM) signal [93]. It is essential to recognize that these energy management strategies have their own set of drawbacks, including limited adaptability to unexpected environmental changes, varying load types, and the complex mathematical computations they require. Specifically, Aziz et al. explore energy management in a PV-battery-diesel hybrid system [94]. Similarly, the possibility of coordinating load tracking and cycle charging through a unified dispatch control strategy for a PV-battery-diesel hybrid system is analyzed by Emara et al. [95]. The research presents a novel direct current (DC) electrical microgrid framework with an advanced energy management system that utilizes PV modules, batteries, and DC loads [95]. Another adaptive dynamics-based EMS for a series-hybrid electric vehicle has been developed in a study by Wang et al. [96]. This approach effectively meets the vehicle's electrical power needs, maintains battery charge levels, and reduces fuel consumption [96]. Additionally, Dai et al. introduced a novel energy management approach that utilizes a wireless sensor network, where evaluating energy needs through data fusion is significant [97]. Gharibeh et al. present a fuzzy logic control-based online multi-mode management system for electric vehicle applications, aiming to reduce battery power fluctuations and increase energy efficiency [98]. Recent research by Ahmed et al. [99], Gaviria et al. [100], and Berghout et al. [101] focuses on a dynamic, integrated microgrid that uses a DC/DC multi-source converter to combine photovoltaics, wind, and hybrid energy storage as renewable energy sources. To operate in various modes, the system employs a power-sharing strategy and a control structure adapted from a standard regulator [99–101].

2.14 Discussion

Intelligent control systems have been extensively researched for various applications, including conventional ventilation control in buildings, power management in photovoltaic systems, and intelligent control systems for current cycles [102–104]. These studies have demonstrated the potential benefits of utilizing artificial intelligence and other control strategies to enhance performance and efficiency in various systems. However, no existing research applies advanced control techniques to coordinate hybrid PV-battery-supercapacitor systems. Our contribution lies in developing an intelligent control framework designed to optimize the performance of hybrid PV-battery-supercapacitor systems using artificial intelligence techniques. The Adaptive Neuro-Fuzzy Inference System (ANFIS) is employed as an intelligent method to control the boost converter, optimizing power delivery to the DC bus. ANFIS is a robust control strategy that combines the adaptability of neural networks with the interpretability of fuzzy logic. Advanced algorithms are utilized in this system to enhance the boost converter's operation and maximize power extraction from the photovoltaic source [105–107].

In conclusion, integrating supercapacitors and batteries into an energy storage system offers numerous advantages. It enhances system efficiency, extends battery lifespan, improves power performance, and increases overall energy density. This integrated approach optimizes energy storage and delivery, enabling greater storage capacity, meeting peak power demands, reducing battery stress, and minimizing energy losses. By leveraging the strengths of both technologies, the integrated system provides a more reliable, efficient, and durable solution for various energy storage applications.

Various configurations of Hybrid Energy Storage Systems (HESS) in PV power generation are commonly simulated using software such as MATLAB/Simulink, HOMER, and P-Graph Studio. Many studies and demonstration projects are currently underway in various parts of the world. Limited information is available on these initiatives due to restricted disclosure by the companies and governments involved. The majority of HESS applications in real-world PV power generation focus on frequency control and backup power integration [53, 108, 109].

In traditional energy management systems for coordinating Primary Battery Voltage (PBV) batteries and supercapacitors, conventional controllers, such as Proportional-Integral-Derivative (PID) controllers, are commonly employed. These controllers are widely used in various control applications due to their simplicity and effectiveness in regulating system variables. The core concept of this configuration is to leverage the strengths of each component: the battery provides short-term support for the low-frequency power component, while the supercapacitor, relying on conventional controllers, offers temporary support for the high-frequency power component. However, these control strategies for energy storage system (ESS) configurations have limitations. They may be unstable under worst-case scenarios, involve complex control algorithms, and require extensive mathematical computations.

This study proposes an intelligent control strategy for coordinating energy storage sources to overcome these limitations. The improved control strategy for the ESS configuration is based on intelligent control methods utilizing machine learning. This approach aims to simplify control algorithms, reduce computational complexity, and enhance the system's overall robustness. By incorporating supercapacitors (SCs) and lithium-ion batteries into the ESS configuration, the intelligent control strategy effectively manages power fluctuations across both low- and high-frequency ranges. Compared to existing approaches, this design offers greater adaptability, enhanced performance, and improved utilization of both energy storage components. This approach enables efficient and practical management of energy flow between system components, leading to improved system performance and reduced energy costs. These benefits enhance reliability, reduce energy costs, and improve system efficiency.

The proposed strategy, which combines intelligent control methods with machine learning techniques, provides a novel approach to improving the performance of grid-connected hybrid PV-battery-supercapacitor systems. This approach may offer significant advantages over current methods, which typically rely on standalone machine learning applications or conventional control strategies. Existing literature has not extensively explored the application of intelligent control strategies for grid-connected hybrid PV-battery-supercapacitor systems.

2.15 Conclusion of the Literature

The proposed system addresses the limitations of existing energy storage control systems by introducing an innovative solution that integrates supercapacitors and lithium-ion batteries. To overcome the current shortcomings of lithium-ion batteries, the Enhanced Energy Storage System employs a control mechanism that separates power components for superior performance. The system's adaptability is further improved by incorporating intelligent artificial intelligence controllers specifically designed to manage switching issues and ensure reliable power delivery despite unexpected variations. Additionally, the seamless integration of the Hybrid Energy Storage System (HESS) into the electrical grid through DQ (D-Q) control transformation regulators enables efficient power flow management, grid synchronization, and voltage and frequency stability. This integrated approach supports bidirectional power flow, allowing the HESS to absorb excess energy during low-demand periods and supply energy back to the grid during peak demand, thereby establishing a flexible and reliable energy management framework. Finally, the intelligent control methods and seamless system integration of the proposed HESS design demonstrate its potential for efficient and effective energy utilization across various scenarios, representing a significant advancement in addressing the challenges associated with power fluctuations.

3 Methodology

3.1 Introduction

The core focus centers on the design and evaluation of a photovoltaic (PV)-battery-supercapacitor (SC) system featuring intelligent energy management and grid connectivity. The study employs MATLAB/SIMULINK, a robust electrical engineering simulation tool, to model and simulate the proposed framework. Due to their potential to enhance the reliability and efficiency of renewable energy systems, the integration of PV arrays and energy storage solutions has garnered considerable attention in recent years. By incorporating a battery-SC hybrid into a solar photovoltaic (PV) system, excess energy generated during peak sunlight hours can be stored and deployed during periods of minimal or absent sunlight, thereby ensuring a consistent and reliable power supply.

The intelligent energy management system is critical in maximizing the performance of the PV-battery-SC system. Furthermore, the simulation will cover crucial parameters and assumptions, along with the performance metrics utilized for evaluating the system's performance. In addition, Fig. 20 presents the research workflow.

3.2 System Topology

Figure 21 depicts the structure of the grid-connected mini-DC system. In order to maximize the amount of PV power extracted, a boost chopper is utilized, which is controlled by a sophisticated MPPT algorithm. It is important to mention that the MPPT technique utilized in this study is fuzzy logic control (FCL). An Artificial Neural Network controls a bidirectional converter that enables the transfer of electrical energy between the DC bus and two energy storage devices, namely batteries and supercapacitors. The DC system is related to the organization through a DC/AC converter. This is followed by an LCL idle channel, assessed to fulfill the rule set by the Association of Electrical and Devices Planners (IEEE1547), as represented in the 2003 arrival of the standard. Based on a few tests [4, 110], researchers have seen that the LCL channel, when used with a standard inverter, is the most incredible in filtering through high-repeat music (HF) delivered by the inverter. This channel beats a direct L or LC channel concerning capability. The utilization of the Phase-Locked Loop (PLL) for synchronizing the system with the power grid [111].

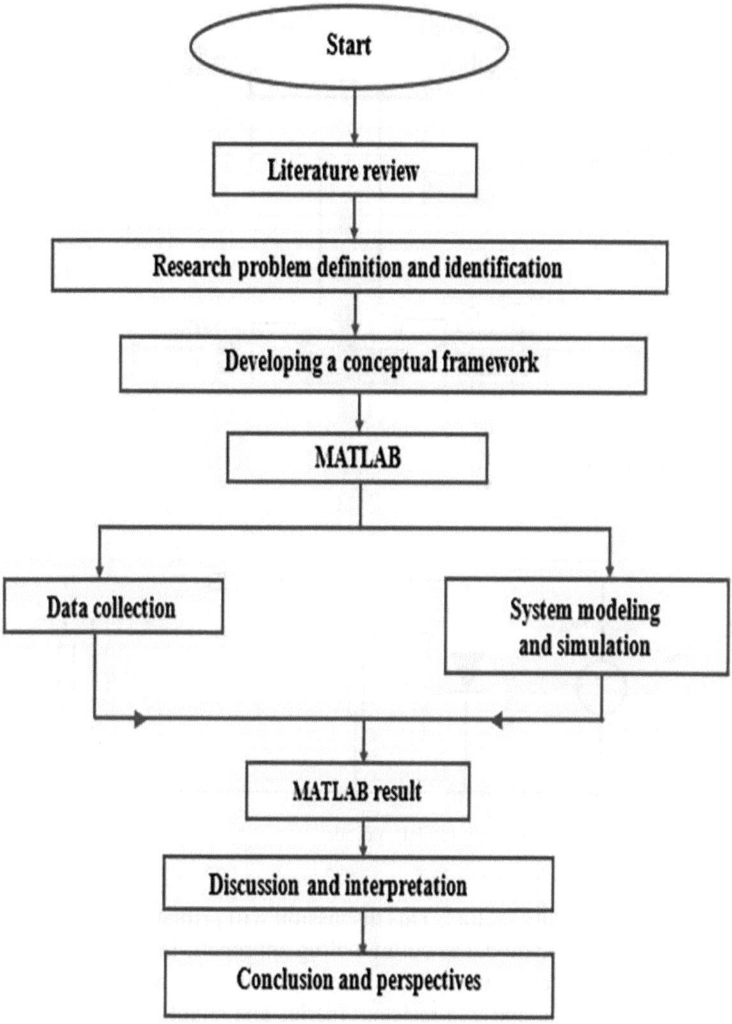

Fig. 20 Research workflow

3.3 *PV Model*

A PV generator typically consists of one or more PV modules connected in series or parallel [112]. It is vital to numerically model the photovoltaic cell to dissect and work on the exhibition of a PV framework successfully. In the past, numerous researchers have developed a variety of models to improve PV systems' suboptimal efficiency as a direct result of this [113]. These mathematical models rely upon speculative circumstances and incorporate concluding different dark cutoff points like the number of diodes, limited or ceaseless shunt obstruction, and unsurprising

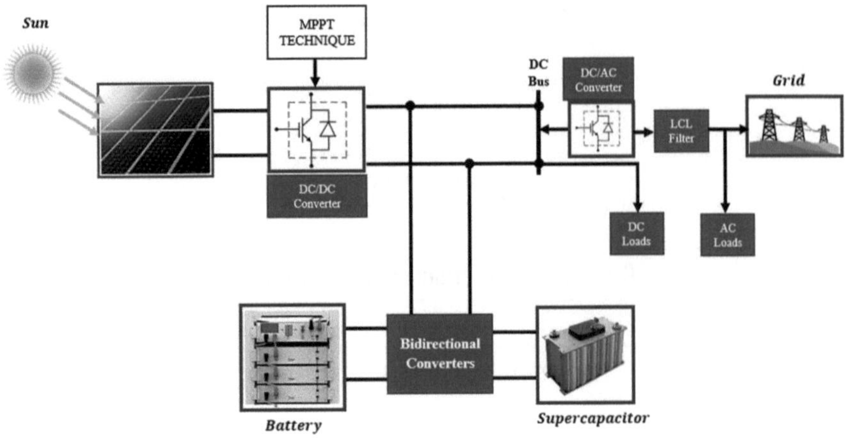

Fig. 21 System topology

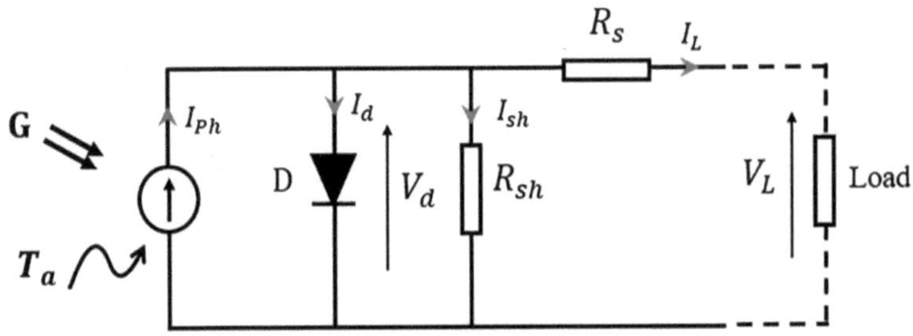

Fig. 22 Equivalent diagram of a cell, models a diode

or non-predictable ideality factors. This discussion will primarily concentrate on the widely used single-diode model as our modeling approach. This model incorporates a current generator to represent the incident luminous flux, a diode to capture cell polarization phenomena, and two resistors (series and shunt) to account for losses. Figure 22 provides a visual representation of the cell model employed in our study.

3.4 Lithium-Ion Battery Model

Battery modeling is difficult because of the complexity of internal chemical reactions. As a result, it is necessary to use a simplified model that can accurately represent the main phenomena occurring inside the cell under specific operating conditions. It is critical to develop a model that is tailored to the behavior of the Li-ion battery based on its intended use. There are several modeling methods available, and the purpose of the application, as well as the methodology used to create the model, allows for differentiation between them. A wide range of modeling perspectives are available. There are various popular Li-ion battery models. Authors show that electrical models based on equivalent electrical circuits are better suited to this application than other models [114].

The design of equivalent circuit models is simple, and their composition varies greatly depending on the performance and accuracy required for the specific application [115].

3.5 Equivalent Circuit Model (ECM)

ECM comprises an ideal voltage source and an Inside resistor (R_{int}). When the battery does not deliver any current, the voltage V_B is equal to the open circuit voltage OCV. When the battery is charging, the current is positive, but when it is discharged, it is negative.

3.6 Thevenin Electric Model

Researchers have created a number of battery models with varying degrees of complexity. As a result, this report uses a first-order model that is well-known for its superior accuracy and simplicity [116, 117]. This battery model is made up of a voltage source that represents the open-circuit voltage (OCV), an ohmic resistor that represents internal resistance, and an RC network that simulates polarization phenomena. The RC network is made up of the capacitance C_{df}, which reflects the polarization of the metal electrodes, and the resistance R_{df}. , which is caused by the electrodes' contact with the electrolyte [118]. This model is shown in Fig. 23, with I-bat representing the charging or discharging current and V_bat representing the voltage at the battery terminal. The formulas are shown in (1).

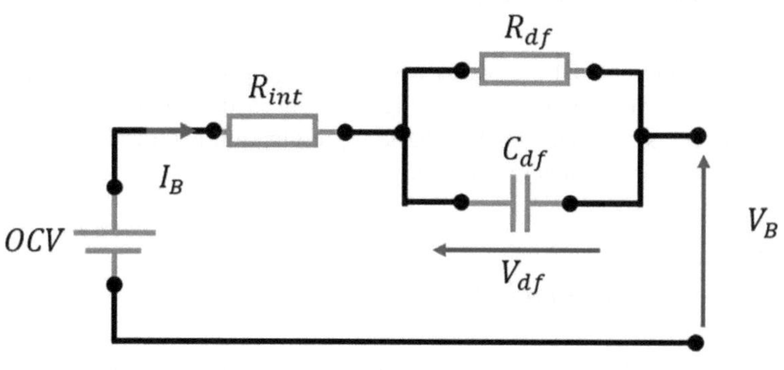

Fig. 23 First-order battery model

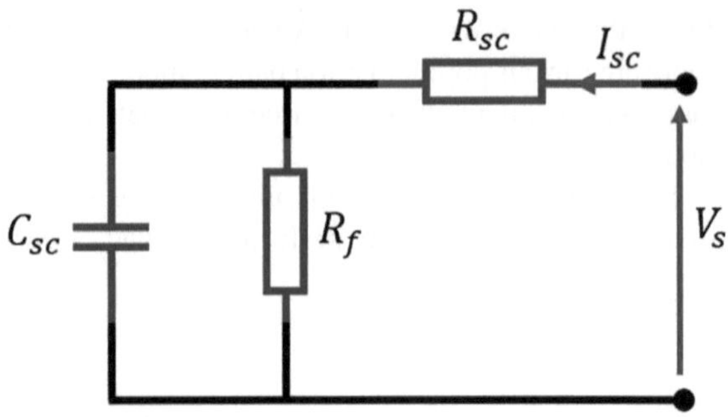

Fig. 24 RC model of the SC

$$\begin{cases} V_B = OCV - RB_{\mathrm{d}f\,\mathrm{int}} \\ V_{\mathrm{d}f} = \frac{I_B}{C_{\mathrm{d}f}} - \frac{V_{\mathrm{d}f}}{R_{\mathrm{d}f} C_{\mathrm{d}f}} \end{cases} \quad (1)$$

3.7 Supercapacity Model

The supercapacitor is the second component of our framework. Supercapacitors are seen as promising developments with positive properties. During the charge cycle, electrically charged particles are moved from the electrolyte to a cathode with the opposite furthest point [119]. The two-branch models, specifically the primary and slow branches, are incorporated into the proposed investigation [120]. In light of the power status of the multi-source framework, this capacity framework's powerful thickness considers quick charging or releasing.

3.8 RC Model of Supercapacitors

SCs can be represented by a series resistance R_{sc}, a escape struggle R_f, and a storage condenser C_{sc}. As observed in Fig. 24, with R_f describing the component's behavior during self-discharge. This basic representation is crucial. It provides an initial understanding of supercapacitors.

3.9 Two-Branch Model of Supercapacitors

The RC two-branch model describes system behavior by breaking down the last response into multiple parts. Each part represents a different constant time.

The Canadians Bonert and Zubieta developed this model, which includes:

- Leakage resistance.
- Two branches with nonlinear capacity and different voltages [121, 122]. Figure 25 displays the supercapacitor cell's equivalent circuit.

Zubieta and Bonert distributed the comeback of a SC into 2 cells:

- The fast branch denotes the supercapacitor's instantaneous response in terms of seconds during a charge or discharge event. Equation (2) expresses this capacity in positions of the energy among its stations, V_1 [83].

$$C_1 = C_0 + C_v V_1 \tag{2}$$

where V_1 is the voltage of C_1

- The second cell is the slow branch, representing charge redistribution during the rest phase. Without considering leakage current, the voltage across the SC can be described by the following equation [83].

$$U_S = N_{s-\mathrm{sc}} V_{\mathrm{sc}} = N_{s-\mathrm{sc}} \left(V_1 + R_1 \frac{I_{\mathrm{sc}}}{N_{\mathrm{Psc}}} \right) \tag{3}$$

where $N_{S-\mathrm{s}}$ and N_{Psc} are the number of supercapacitor cells in series and equivalent affiliation, independently. Similarly, I_sc is the supercapacitor module's current, and V_S is the cell voltage.

For the slow cell, the voltage V2 can be expressed as [83]:

$$V_2 = \frac{1}{C_2} \int I_2 \mathrm{d}t = \frac{1}{C_2} \int \frac{1}{R_2} (V_1 - V_2) \mathrm{d}t \tag{4}$$

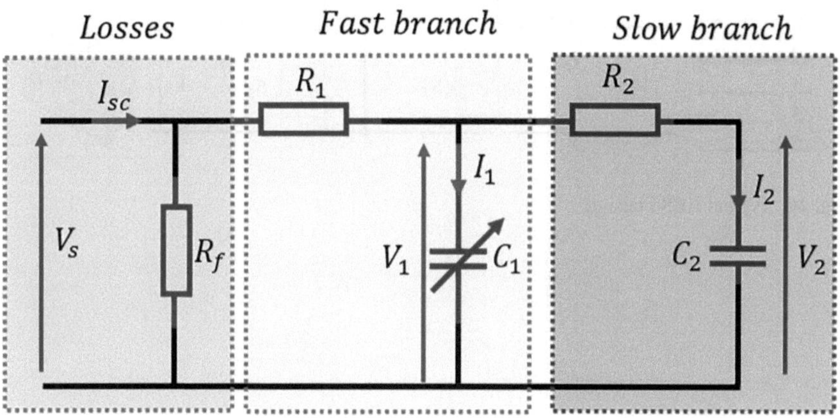

Fig. 25 RC model of the supercapacitor

3.10 DC/DC Converter Model

Through bidirectional DC/DC converters, Hybrid Energy Storage Systems (HESS) are connected to the DC network. HESS keeps a predictable DC structure voltage V_0 due to a mismatch between age and solicitation. This work represents the total load of Remote Area Power System Applications (RAPSA) as an equivalent direct current load with resistance R. Figure 26 represents HESS topologies.

Let's denote the voltages of the PV panel, battery, and solar cell as V_{pv}, V_B, and V_s, respectively. Correspondingly, the currents from the PV panel, battery, and solar cell are represented by I_{pv}, I_B, and I_S. The filter inductance for boost converters, batteries, and SC converters are denoted by L_{pv}, L_B, and L_S, respectively. The dc grid voltage is represented by V_0, while C stands for the filter capacitance and R denotes the load resistance. Control switches are labeled as S_1, S_2, S_3, S_4, and S_5 [110].

A boost converter connects the solar generator to the direct current grid. The maximum power output is controlled using the MPPT algorithm.

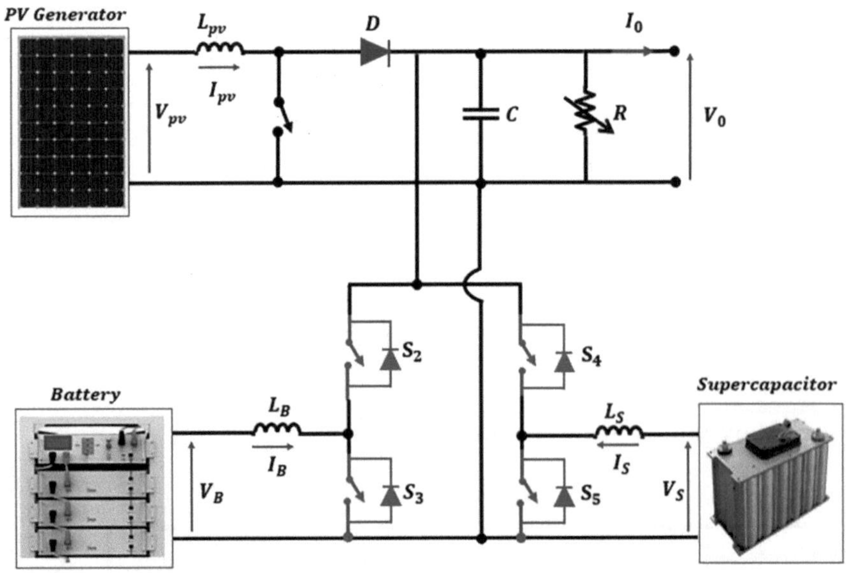

Fig. 26 Typical HESS design

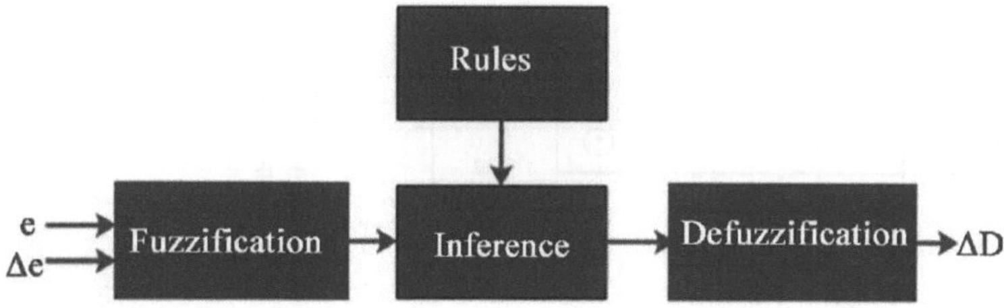

Fig. 27 Diagram of the FLC block [124]

3.11 MPPT Technique

Various MPPT techniques are documented in the literature, each designed to extract the maximum power from PV modules across diverse applications. These techniques vary in their effectiveness, complexity, sensor requirements, ease of hardware implementation, convergence speed, and other factors [4].

Irrespective, as opposed to the commonly used perturb and observe (P&O) computation, a fuzzy logic-based controller is utilized in light of its better speed and execution performance than various estimations. Fuzzy logic controllers (FLCs) can work with loose data sources, are separate from exact numerical models, and are effective in managing nonlinearities. The FLC is used to direct the PV generator [123]. Indeed, the FLC modules are divided into three categories:

- fuzzification,
- decision-making, and
- defuzzification.

The architecture of the FLC block is shown in Fig. 27 [124]. E, CE, and liability series signify the contribution and production of the FLC.

3.12 LCL Filter Model

The inverter is the primary power electronic converter that facilitates the connection of renewable energy sources to the utility grid. Regardless, the high-repeat trading of semiconductors could impel symphonious curving in the system. To mitigate this, the grid filter is employed to reduce harmonic injection into the utility grid.

The principal support for picking the LCL channel is that it will, in general, be used at low trading frequencies, has greater channel viewpoints than ordinary "L" and "LC" channels, has a lower voltage drop, and has better damping [125].

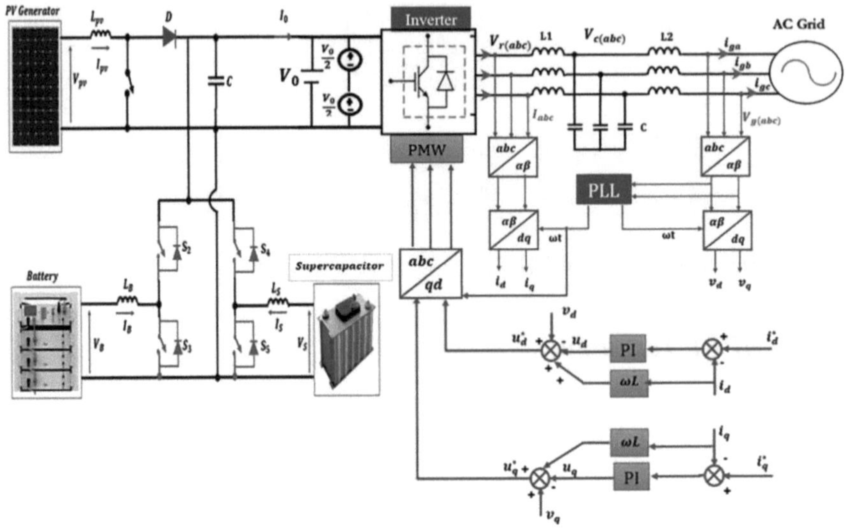

Fig. 28 System block diagram with DQ control

3.13 DQ Control

DQ control of inverter connected to grid is a control strategy used in power electronics and electrical engineering. In this method, the system's currents and voltages are converted from the three-phase (abc) to the DQ reference frames. "D" denotes the direct axis, which is parallel to the inverter's DC link voltage, and "Q" denotes the quadrature axis, which is perpendicular to the D-axis. Control becomes more manageable by transforming the variables into this reference frame, particularly for grid-connected systems such as inverters. The DQ coordinate system allows for non-static error control of grid-connected current using the classical PI controller. This method is flexible and simple to implement, making it widely used [126, 127]. Figure 28 shows the control of the system.

3.14 Conventional Energy Management Strategy for the HESS

Figure 29 describes the conventional strategy [128]. The objective of this computation is to restrict the charging and delivering loads on the lithium-molecule battery, consequently overhauling its life expectancy. In this assessment, the normal DC system voltage (V_O) is differentiated from the reference voltage (V_{ref}), with the PI controller getting the ensuing misstep. The PI controller pertaining to the energy storage system (ESS) computes the total required current.

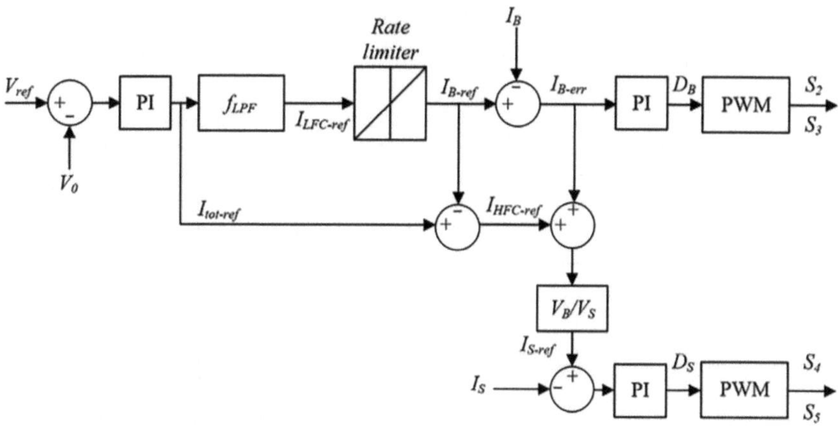

Fig. 29 Conventional HESS strategy scheme [128]

3.15 Proposed System Management Strategy for the HESS

Figure 30 depicts the block diagram for the proposed HESS control strategy. This proposed strategy represents an advancement over the approach introduced in [129]. The PI controller plays a major role in controlling switch commutation in the traditional approach. However, the PI controller has drawbacks, including a large initial overshot, a slow reaction to unanticipated disturbances, and a vulnerability to controller gains. The suggested system responds well to sudden variations in load and climate while maintaining a significant power flow. It is recommended that intelligent controllers be used to address problems that are frequently connected to classical controllers, particularly those that have to do with switching.

The proposed HESS control strategy emphasizes the use of ANN the conventional energy management strategy or of Sliding Mode (SM) controller units, which offer greater robustness compared to traditional controllers. They are adept at handling sudden variations, thereby enhancing the performance and lifespan of the storage system while also averting load disconnection. This study primarily focuses on machine learning, with a brief overview of SM control provided.

3.16 SM Controller

In [128], the energy management system incorporates effective control mechanisms, employing both ANN and SM controllers. The SM controller ensures precise control of the SC bidirectional converter, while the ANN controller governs the operation of the remaining bidirectional converter. VSC and their sliding regimes (referred to as Sliding Modes (SM)) are distinguished by a control discontinuous aspect that specifies a desirable system dynamic. Such dynamics are achieved by selecting appropriate

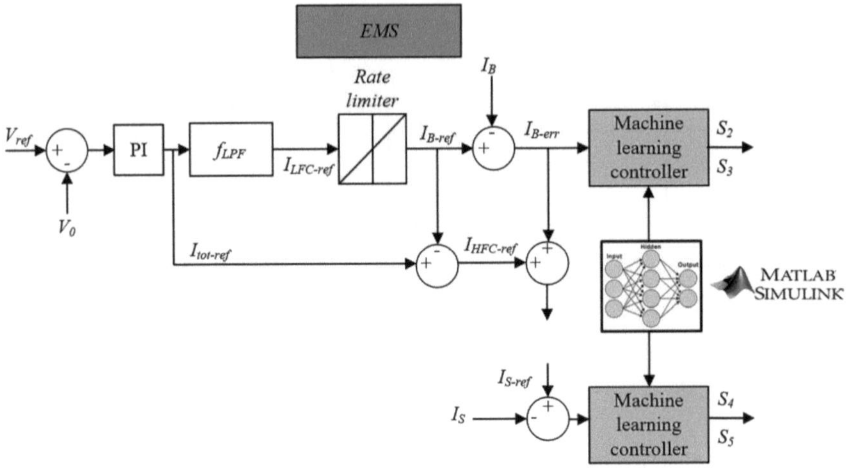

Fig. 30 Proposed HESS strategy scheme

variety spaces, also known as sliding surface controllers, which provide adequate switching and result in the desired behavior of closed loop systems.

3.17 Machine Learning

Machine learning (ML) is a subfield of programming and Artificial Intelligence that teaches computers to think exactly like people do. PC-based information evaluations use PC enhancements to "learn" information clearly from data rather than include an earlier condition as a model. The calculations change and work on their showcase as the quantity of accessible learning models develops. Individual learning (bunching) reveals hidden examples or intrinsic designs in the input information. In contrast, regulated learning (classification and regression) prepares a model based on known information and result information to predict future results [130].

Picking the real estimation can be overpowering as there are many coordinated and solo Artificial Intelligence computations, each with a clever method for managing learning. There is no one-size-fits-all approach or method. Finding the genuine computation is generally established on trial and error; even incredibly experienced data specialists can't predict expecting estimation will work without testing it [130].

3.18 The ANN Controller

This research introduces an energy management approach that exclusively utilizes ANN controllers. The conventional controller is substituted with an ANN controller,

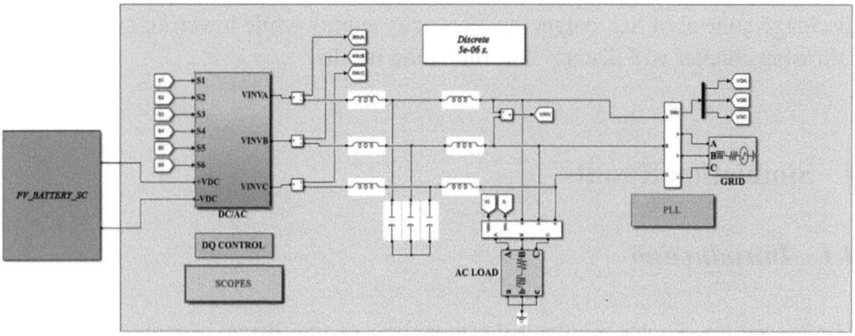

Fig. 31 Advanced strategy simulation using MATLAB/Simulink environment

thereby enabling the use of ANN controllers to regulate each individual converter. The ANN controller has become increasingly favored in energy management strategies [129]. Its architecture comprises three layers: input, hidden, and output, with each layer featuring neurons (also referred to as nodes) interconnected within multilayer networks. The tasks of the ANN controller encompass two stages: training and operational. Artificial Neural Networks (ANNs) are highly suitable for integration into control systems due to their inherent advantages, such as their ability to handle nonlinear and adaptable processes, exhibit strong generalization capabilities, and operate independently of the specific parameters of the system design. The objective of implementing an Artificial Neural Network (ANN) control system within the energy storage system is to optimize power generation by dynamically adjusting the control approach. Through a gradual learning process, the ANN is designed to adapt and effectively utilize available system resources, thereby enhancing the overall efficiency of the energy storage system [131]. The supercapacitor and battery charging and discharging are controlled by two ANN controllers.

Utilizing MATLAB/Simulink, the intended HESS linked to the electrical grid control strategy implementation is managed, as shown in Fig. 31.

3.19 Summary of Section

In conclusion, the suggested technique for developing a PV-battery supercapacity system with intelligent energy management coupled to the grid utilizing MATLAB SIMULINK provides a complete strategy to optimizing solar energy consumption and improving electrical grid stability. The system's performance can be effectively evaluated by accurately modeling and simulating its components, such as PV panels, SC, MPPT techniques, electrical grid, and energy management algorithms. This allows for an assessment of its performance under various operating situations. The integration of intelligent energy management allows for effective charge and discharge control of SC, balancing electricity supply while lowering expenses. The following chapter will discuss the simulation results.

4 Simulation Results

4.1 Introduction

The preceding chapter involved the modeling of the different components of the system. Moving forward, grid-connected Hybrid Energy Storage System (HESS) will be simulated using MATLAB/Simulink, specifically for a one-day load curve. By utilizing the Artificial Neural Network, valuable deductions can be made regarding the suitability of the suggested control strategy.

4.2 Software Used

The utilized MATLAB/Simulink, occasionally known as the Alliance Assessment Office, for this project. MATLAB is frequently utilized by experts in a variety of fields, including computer engineering (CE) and electrical engineering (EE). In MATLAB, there are two related parts. The programming language, which has two or three fundamental programming developments, contains the majority of the components. In keeping with the piece, there are a variety of crucial responsibilities and constraints that are frequently utilized in planning and science spaces. Therefore, MATLAB's scholar and expert achievement will help plan understudies [132].

4.3 Description of the Studied Photovoltaic Conversion Chain Connected to the Grid

When carrying out the present task, a 7kW PV generator is used. In the framework being scrutinized, a bidirectional converter is constrained by an Artificial Neural Network (ANN). It facilitates the transfer of electricity from the DC bus to energy storage devices such as supercapacitors and batteries. A DC/AC converter and an LCL latent channel are utilized to interface the DC matrix to the network. It is noteworthy that in order to verify the superior performance of the advanced architecture, the design is subjected to comparison with several well-known conventional techniques, taking into consideration any unforeseen deviations.

4.4 Simulation Results and Interpretations

This study uses a 7 kW photovoltaic generator. The real climatic conditions profile is shown in Fig. 32.

It is important to note that, as shown in Fig. 33, a significant increase or decrease in the heap current addresses each of the stores high and low groups. The peak demand of energy of the load is between 12h–15h.

Figure 34 shows the filtered waveforms of AC load currents with the designed LCL filter. AC load current has a maximum value of 23A.

Machine learning techniques were employed to implement intelligent control strategies in the system's energy management. This allowed for enhanced decision-making capabilities and optimization of system performance. The error histogram's results are shown in Fig. 35. After conducting machine learning training, validation, and testing, the results demonstrated that the device plans system consistently achieved a zero error with negative values (− 0.01348) in all instances. This demonstrates the effectiveness of ANN controllers over traditional techniques. The studied HESS control's robustness can be preserved by ANN.

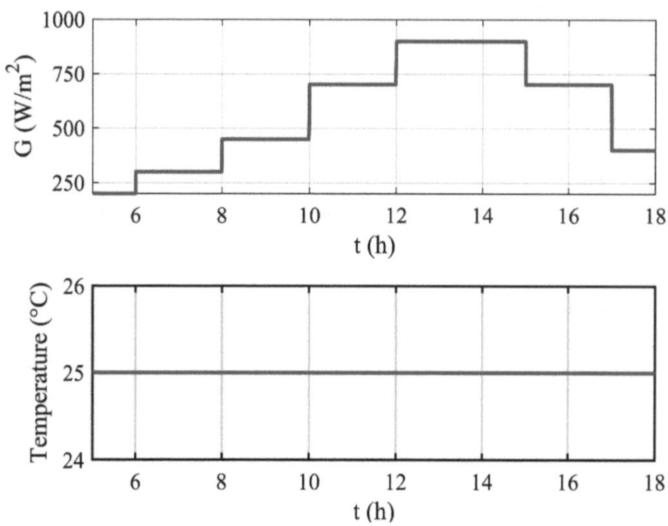

Fig. 32 Real climatic conditions profile

Fig. 33 DC load profile

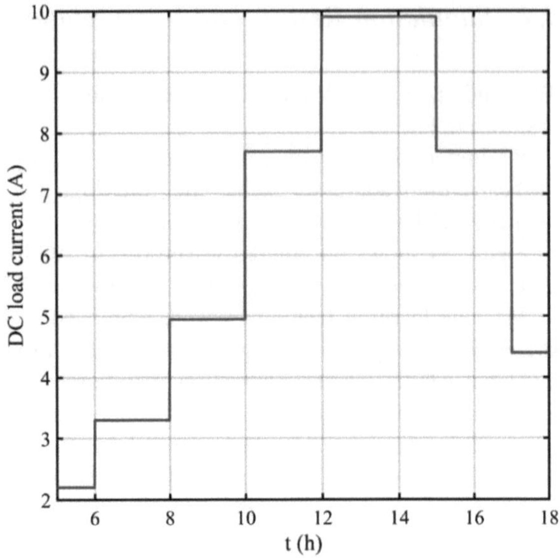

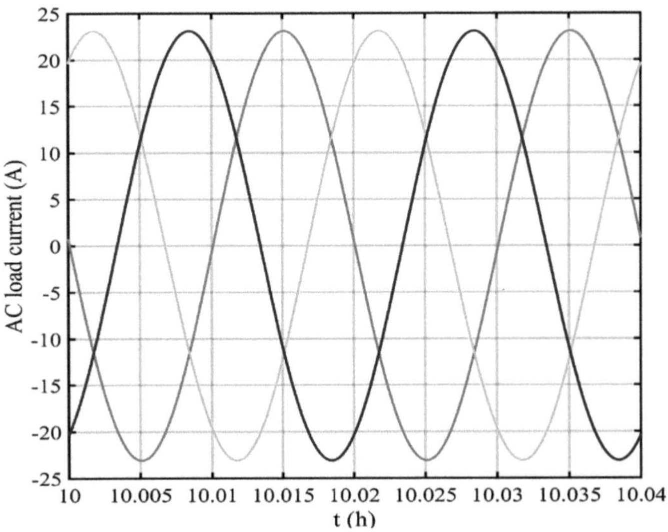

Fig. 34 AC load profile

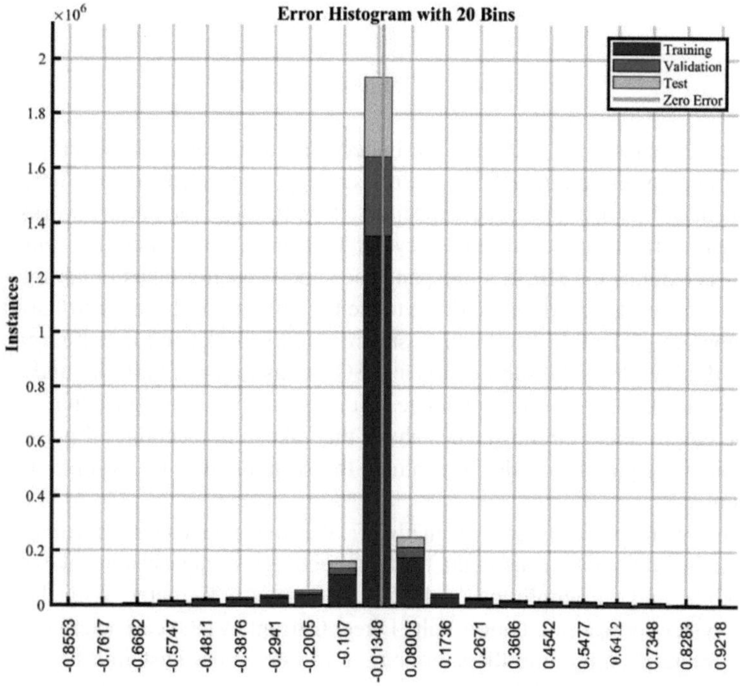

Fig. 35 Error histogram

$$\text{Errors} = \text{Targets} - \text{Outputs}$$

Figure 36 shows the SOC for the battery and supercapacitor separately. Also, Fig. 36 depicts the entire HESS control strategy's approach to the charging and movement of batteries and supercapacitors. True to form, the battery works impeccably during the charge and release cycle. This decrease in operating temperature and stress extends the HESS's lifespan. The supercapacitor takes in abrupt voltage changes. Bringing the lift record down to the DC affiliation where controllability and execution are preferable is related to supercapacitor packs and batteries. Since the models don't address the submersion and SOC constraints of batteries and supercapacitors, they have little effect on the test. It is essential to emphasize that a comprehensive environment circumstances shape that takes into account all possible load variations is required in order to evaluate the sustainability of the optional HESS switch plan.

DC bus voltages were displayed in Fig. 37. The most noticeable overvoltage occurs between the hours of five and eight, the battery energy storage system takes one hour to stabilize, and the amplitude of the fluctuations is not greater. With the help of HESS, system power fluctuation Volts Direct Current (VDC) can be effectively and swiftly reduced, mitigating the shortcoming of the battery system's slow response to sudden changes in load demand. Because machine learning is utilized to stabilize the voltage on the DC bus, there is minimal variance in the DC voltage even when the weather varies.

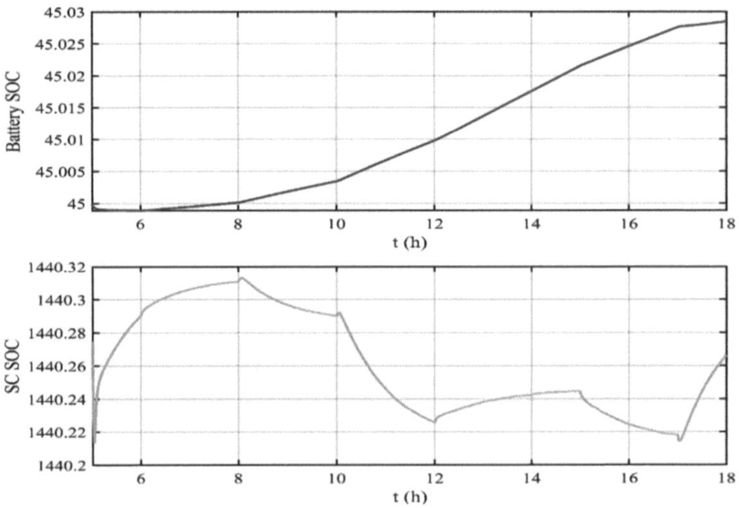

Fig. 36 Superior lithium-ion battery-operated capacity and SOC

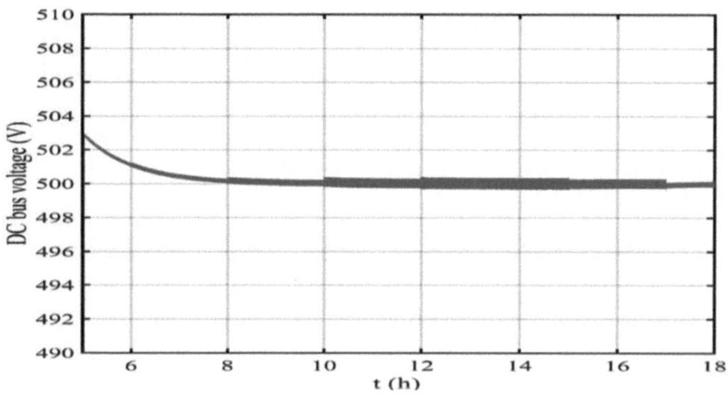

Fig. 37 DC bus voltage

The reactive and active power are shown in Fig. 38. The PLL appears to remain going correctly when the d voltage stabilizes and the q voltage becomes null. Since the infused power into the network is completely dynamic and steady during the administration time spans, Assortments in irradiance don't impact the controllability of the PV plant. The number of bidirectional converter characteristics has increased. The components of the framework were developed with consideration for a variety of lighting conditions in view of a typical day-to-day profile in order to match the ideal behavior. The use of Artificial Neural Network (ANN) and fuzzy Logic control (FLC) control techniques were expected to achieve three essential targets: (1) expanding the power yield from PV sheets through MPPT control; (2) choosing the best errand parting procedure between the battery and SC; and (3) empowering the framework through direct-quadrature hypothesis.

Figure 39 depicts a window of the inverter and framework voltage waveforms with what appears to be power injection. Utilizing a Phase-Locked Loop permits the two parts to be in a similar stage, which is important for framework synchronization with the lattice.

The study demonstrates the validity of the control method by reducing the harmonic distortion rate of the grid-connected current by 1.06% as a result of the PLL's negative influence when compensated on the q-axis. The LCL filter lowers high-degree harmonics in addition to PLL, resulting in a lower total harmonic distortion (THD) (Fig. 40). By employing intelligent control strategies, the system's total THD (1.06%) is less than the 5% recommended by IEEE standards [131].

4.5 Summary of the Section

The goal of this investigation was to thoroughly examine a grid-connected mixture renewable energy system that included PV, battery, and SC technologies. Special emphasis was placed on optimizing control and energy management techniques. An

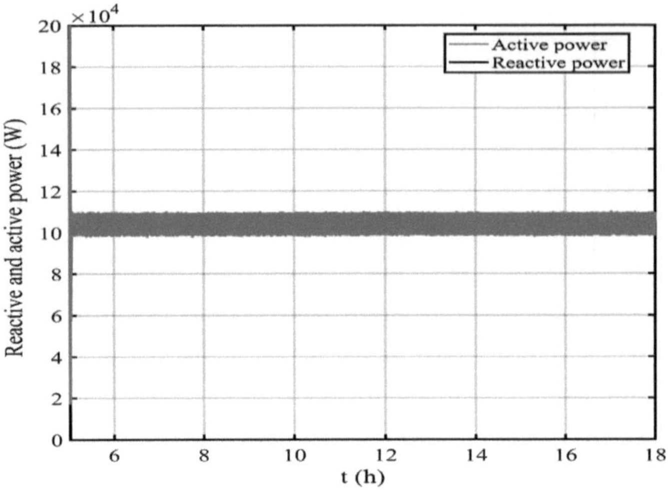

Fig. 38 Reactive and active power

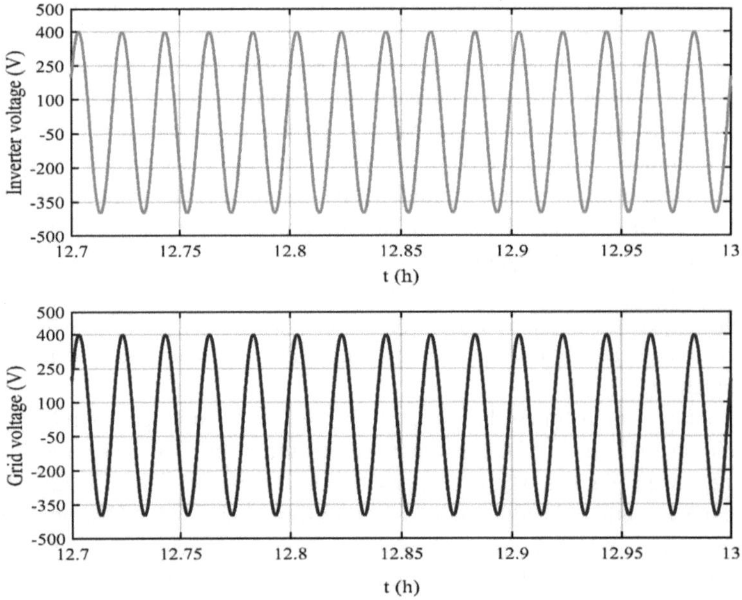

Fig. 39 Inverter and grid voltage

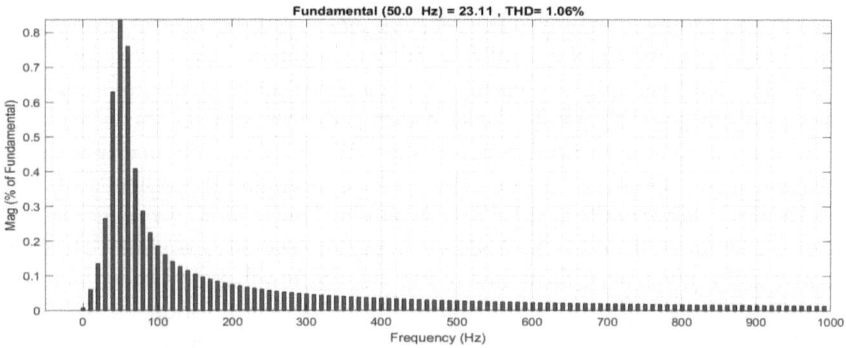

Fig. 40 Total harmonic distortion

innovative management approach based on intelligent techniques has been presented for the grid-connected hybrid green power system. The results of implementing this technique revealed its efficacy, as proven by a low THD value of 1.06% and stable voltage regulation, even under fluctuating climatic and load circumstances. This outstanding accomplishment demonstrates the suggested system's exceptional performance and efficiency when using the recommended management method.

5 Conclusion

The objective of this study's HESS connected to grid-connected control strategy, which uses machine learning techniques to connect grid-connected hybrid PV-battery-supercapacitor systems, is to screen both the high-repeat and low-repeat power portions of the interest age bewilder while utilizing the battery current botch to drop by the best outcomes.

The high-recurrence parts and a piece of the battery current are shipped off the SC. As a result, the charge/discharge current rates are significantly reduced, extending the battery's lifespan and enabling a smaller battery size. Experiments are conducted to validate the suggested architecture and approach for abrupt changes in the reference power to be injected into the grid.

The study effectively integrates machine learning techniques, particularly Artificial Neural Networks (ANNs) and fuzzy logic control (FLC), into the energy management of a grid-connected hybrid system powered by a 7kW photovoltaic (PV) generator. The system incorporates a bidirectional converter, constrained by the ANN, facilitating the seamless transfer of electricity between the DC bus and energy storage devices such as supercapacitors and batteries. The DC/AC converter and LCL filter are crucial in interfacing the DC grid with the main power network. Simulation results indicate that the PV generator's power output varied significantly under different weather conditions, with peak energy demand occurring between 12 and 15 h. The use of machine learning strategies enhanced the system's decision-making capabilities, leading to improved performance and stability. The error histogram in Fig. 35 highlights the ANN controller's ability to maintain a near-zero error margin (-0.01348), demonstrating its robustness compared to traditional control techniques. The state of charge (SOC) profiles for the battery and supercapacitor, shown in Fig. 36, reveal that the Hybrid Energy Storage System (HESS) effectively manages sudden voltage changes, reducing operational stress and extending the system's lifespan. The DC bus voltage profile in Fig. 37 shows minimal fluctuation, even under varying weather conditions, thanks to the ANN's voltage stabilization capabilities. The grid's reactive and active power, illustrated in Fig. 38, confirm that the system maintains dynamic and steady power output, unaffected by irradiance variations. The study also demonstrates a reduction in harmonic distortion to 1.06% in the grid-connected current, well below the 5% threshold recommended by IEEE standards, as shown in Fig. 40. This reduction in total harmonic distortion (THD) is attributed to the combined effects of the LCL filter and the Phase-Locked Loop (PLL) compensations. Advanced architecture proves to be significantly more efficient and stable than conventional methods, effectively addressing the challenges posed by fluctuating weather conditions and varying load demands.

The findings demonstrate that the suggested plan can lower the charge/discharge rates and give the current injected into the grid an almost sinusoidal waveform. Utilizing the PV array's power curves has also helped with HESS sizing and battery/supercapacitor distribution. Depending on the converter's nominal power and the time interval that the generator wishes to be managed, a HESS capacity range is computed.

Artificial Neural Network (ANN) controllers are employed to enhance the performance and lifespan of the storage system. They also prevent the load from being completely disconnected, ensuring the robustness of the investigated Hybrid Energy Storage System (HESS) connected to grid-connected control. These controllers consider sudden scenarios of unexpected climatic conditions and load variations. Machine learning decreases the presence of current harmonics. Unlike the existing systems, the proposed methodology demonstrates reduced complexity and enhanced ease of implementation.

When it appears differently in relation to other standard organization structures, the lithium-molecule battery's term cycle is widened, and top streams during charging and delivering are decreased because the low-repeat part is moved to the battery. The recommended HESS, which utilizes a grid-connected control method, has been successfully implemented through the use of MATLAB/Simulink. This implementation involves connecting the FLC and PV generator, the SC, and the lithium-molecule battery modules. The simulation results clearly show that the studied HESS control strategy is superior, efficient, and outperforms conventional strategies in several aspects. These include reducing oscillations, regulating bus voltage in a timely and continuous manner, increasing battery lifespan, and ensuring the continuity and persistence of the PV system.

5.1 Future Works

In terms of future works, there are numerous factors to consider:

- Additional renewable sources of energy, such as wind power, can be considered. This would increase the renewable energy mix and improve the overall system's sustainability.
- It would be good to look at how distribution network design tactics have changed over time in response to advances in smart grid technology. Understanding changes in planning methodologies can help to improve the efficiency and reliability of grid-connected systems.
- Versatile AC Transmission Systems (Real factors) can be used to foster power quality further, especially when the association is working at its full breaking point. Implementing FACTS systems can assist in reducing voltage fluctuations, increase system stability, and boost overall performance.

Declaration of Generative AI and AI-Assisted Technologies in the Writing Process During the preparation of this work the author(s) used Quillbot, and GPT4 of openAI in order to improve the language. After using this tool/service, the author(s) reviewed and edited the content as needed and take(s) full responsibility for the content of the publication.

References

1. Safaei B, Erdem S, Kolamroudi MK, Arman S (2024) State-of-the-art review of energy harvesting applications by using thermoelectric generators. Mech Adv Mater Struct 31(22):5605–5637
2. Kolamroudi MK, Ilkan M, Egelioglu F, Safaei B (2022) Maximization of the output power of low concentrating photovoltaic systems by the application of reflecting mirrors. Renew Energy 189:822–835
3. Hafez FS, Sa'di B, Safa-Gamal M, Taufiq-Yap YH, Alrifaey M, Seyedmahmoudian M, Mekhilef S et al (2023) Energy efficiency in sustainable buildings: a systematic review with taxonomy, challenges, motivations, methodological aspects, recommendations, and pathways for future research. Energy Strat Rev 45:101013
4. Mokhlis M, Ferfra M, Vall HA, Ahmed CC, Taouni A (2020) Comparative study between the different MPPT techniques. In: Proceedings of the 5th international conference on renewable energies. IEEE
5. Nkambule MS, Hasan AN, Ali A, Hong J, Geem ZW (2021) Comprehensive evaluation of machine learning MPPT algorithms for a PV system under different weather conditions. J Electr Eng Technol 16(1):411–427
6. Venkateswaran R, Natesan B, Lee SR, Joo YH (2022) Maximum power extraction for PMVG-based WECS using Q-learning MPPT algorithm with finite-time control scheme. IEEE Trans Sustain Energy
7. Guan Z, Yamamoto T (2021) Design of a reinforcement learning PID controller. IEEJ Trans Electr Electron Eng 16(10):1354–1360
8. Batiyah S, Sharma R, Abdelwahed S, Alhosaini W, Aldosari O (2022) Predictive control of PV/battery system under load and environmental uncertainty. Energies 15(11):4100

9. Rico-Riveros LF, Trujillo-Rodríguez CL, Díaz-Aldana NL, Rus-Casas C (2023) Modelling of electric power generation plant based on gas turbines with agricultural biomass fuel. Electronics 12(9):1981
10. Roy D, Samanta S, Ghosh S (2020) Performance assessment of a biomass fuelled advanced hybrid power generation system. Renew Energy 162:152–164
11. Sharma S, Jain P (2020) Integrated TOU price-based demand response and dynamic grid-to-vehicle charge scheduling of electric vehicle aggregator to support grid stability. Int Trans Electr Energy Syst 30(1):e12160
12. Wu J, Zhang Q, Lu Y, Qin T, Bai J (2023) Source-load coordinated low-carbon economic dispatch of microgrid including electric vehicles. Sustainability 15(21):15287
13. Yuan Q, Ye Y, Tang Y, Liu X, Tian Q (2022) Low carbon electric vehicle charging coordination in coupled transportation and power networks. IEEE Trans Ind Appl
14. Ren Y, Rind SJ, Jiang L (2020) A coordinated control strategy for battery/supercapacitor hybrid energy storage system to eliminate unbalanced voltage in a standalone AC microgrid. J Intell Manuf Spec Equipment 1(1):3–23
15. Kolamroudi MK, Ilkan M, Egelioglu F, Safaei B (2023) A comparative study of LCPV by mirror reflection against other systems: recent techniques, implications, and performances. Solar Energy 247:1–15
16. Punna S, Banka S, Salkuti SR (2022) Optimal energy management scheme of battery supercapacitor-based bidirectional converter for DC microgrid applications. Information 13(7):350
17. Chiras DD (2017) Power from the sun: achieving energy independence. New Society Publishers
18. Basit MA, Dilshad S, Badar R, Sami ur Rehman SM (2020) Limitations, challenges, and solution approaches in grid-connected renewable energy systems. Int J Energy Res 44(6):4132–4162
19. Hamid AK, Mbungu NT, Elnady A, Bansal RC, Ismail AA, AlShabi MA (2023) A systematic review of grid-connected photovoltaic and photovoltaic/thermal systems: benefits, challenges and mitigation. Energy Environ 34(7):2775–2814
20. Khezri R, Mahmoudi A (2020) Review on the state-of-the-art multi-objective optimisation of hybrid standalone/grid-connected energy systems. IET Gener Transm Distrib 14(20):4285–4300
21. Jain S, Sawle Y, Zubair M, Khalil M (2021) Optimization and comparative economic analysis of standalone and grid-connected hybrid renewable energy system for remote location. Front Energy Res 9:724162
22. Morey M, Gupta N, Garg M, Kumar A (2023) A comprehensive review of grid-connected solar photovoltaic system: architecture, control, and ancillary services. Renew Energy Focus 46:100–120
23. Elalfy DA, Gouda E, Kotb MF, Bureš V, Sedhom BE (2024) Comprehensive review of energy storage systems technologies, objectives, challenges, and future trends. Energy Strat Rev 54:101482
24. Singh P, Lather JS (2020) Dynamic power management and control for low voltage DC microgrid with hybrid energy storage system using hybrid bat search algorithm and artificial intelligence. J Energy Storage 32:101879
25. Liu Y, Xu Z, Hua L, Zhao X (2020) Analysis of energy characteristic and working performance of novel controllable hydraulic accumulator with simulation and experimental methods. Energy Convers Manag 222:113238
26. Saha P, Zhao W, Stroe DI, Iov F, Munk-Nielsen S (2023) Enabling LVRT compliance of electrolyzer systems using energy storage technologies. Batteries 9(11):527
27. Soomar AM, Hakeem A, Messaoudi M, Musznicki P, Iqbal A, Czapp S (2022) Solar photovoltaic energy optimization and challenges. Front Energy Res 10:879985
28. Kruitwagen L, Story KT, Friedrich J, Byers L, Skillman S, Hepburn C (2021) A global inventory of photovoltaic solar energy generating units. Nature 598(7882):604–610

29. Sayed ET, Olabi AG, Alami AH, Radwan A, Mdallal A, Rezk A, Abdelkareem MA (2023) Renewable energy and energy storage systems. Energies 16(3):1415
30. Candra O, Chammam A, Alvarez J, Martínez I (2023) The impact of renewable energy sources on the sustainable development of the economy and greenhouse gas emissions. Sustainability 15(3):2104
31. Chien F, Hsu C, Ozturk I, Sharif A, Sadiq M (2022) The role of renewable energy and urbanization towards greenhouse gas emissions in top Asian countries: evidence from advanced panel estimations. Renew Energy 186:597–609
32. Li X, Raorane CJ, Xia C, Wu Y, Tran TKN, Khademi T (2023) Latest approaches on green hydrogen as a potential source of renewable energy towards sustainable energy: spotlighting of recent innovations, challenges, and future insights. Fuel 334:126684
33. Sharma G, Shah M, Shahzad U, Jain M (2021) Exploring the nexus between agriculture and greenhouse gas emissions in BIMSTEC region. J Environ Manag 299:113587
34. Das S, Kumar J, Dawn S, Singh F (2023) Existing stature and possible outlook of renewable power in comprehensive electricity market. Processes 11(6):1849
35. Prasad S, Sheetal KR, Venkatramanan V, Kumar S, Kannojia S (2019) Sustainable energy: challenges and perspectives. In: Sustainable green technologies for environmental management, pp 175–197
36. Mendis N, Muttaqi KM, Perera S (2012) Active power management of a super capacitor-battery hybrid energy storage system for standalone operation of DFIG based wind turbines. In: 2012 IEEE industry applications society annual meeting. IEEE, pp 1–8
37. Sathishkumar R, Kollimalla SK, Mishra MK (2012) Dynamic energy management of micro grids using battery super capacitor combined storage. In: 2012 annual IEEE India conference (INDICON). IEEE, pp. 1078–1083
38. Glavin ME, Chan PK, Armstrong S, Hurley WG (2008, September) A stand-alone photovoltaic supercapacitor battery hybrid energy storage system. In: 2008 13th international power electronics and motion control conference. IEEE, pp 1688–1695
39. Şahin ME, Blaabjerg F (2020) A hybrid PV-battery/supercapacitor system and control proposal in MATLAB/Simulink. Electronics 9(1):129
40. Liu W, Liu X, Du B, Jiang Z, Cui H, Gao Q, Wei Y, Li N, Xi D, Liu C (2016) Research on hybrid energy storage by super capacitor-battery and control strategy for intelligence microgrid system. In 2016 IEEE 2nd annual Southern power electronics conference (SPEC). IEEE, pp 1–5
41. Bharatee A, Ray PK, Subudhi B, Ghosh A (2022) Power management strategies in hybrid energy storage systems. Energies 15(19):7176
42. Liu Z, Du J, Yu B (2020) Design method of double-boost DC/DC converter with high voltage gain for electric vehicles. World Electr Veh J 11(4):64
43. Ali AIM, Alaas ZM, Sayed MA, Almalaq A, Farah A, Mohamed MA (2022) An efficient MPPT technique-based single-stage incremental conductance for integrated PV systems considering flyback central-type PV inverter. Sustainability 14(19):12105
44. Remoaldo D, Jesus I (2021) Analysis of a traditional and a fuzzy logic enhanced perturb and observe algorithm for the MPPT of a photovoltaic system. Algorithms 14(1):24
45. Khan MJ, Mathew L, Alotaibi MA, Malik H, Nassar ME (2022) Fuzzy-logic-based comparative analysis of different maximum power point tracking controllers for hybrid renewal energy systems. Mathematics 10(3):529
46. Khan MA, Abbas K, Su'ud MM, Salameh AA, Alam MM, Aman N, Mehreen M, Jan A, Hashim NAABN, Aziz RC (2022) Application of machine learning algorithms for sustainable business management based on macro-economic data: supervised learning techniques approach. Sustainability 14(16):9964
47. Phan BC, Lai YC, Lin CE (2020) A deep reinforcement learning-based MPPT control for PV systems under partial shading condition. Sensors 20(11):3039
48. Abdel-Rahim O, Chub A, Blinov A, Vinnikov D, Peftitsis D (2022) An efficient non-inverting buck-boost converter with improved step up/down ability. Energies 15(13):4550

49. Sayed K, Almutairi A, Albagami N, Alrumayh O, Abo-Khalil AG, Saleeb H (2022) A review of DC-AC converters for electric vehicle applications. Energies 15(3):1241
50. Nguyen TVH, Pham CK (2025) An overview of phase-locked loop: from fundamentals to the frontier. Sensors (Basel, Switzerland) 25(18):5623
51. Adamas-Pérez H, Ponce-Silva M, Mina-Antonio JD, Claudio-Sánchez A, Rodríguez-Benítez O, Rodríguez-Benítez OM (2024) A new LCL filter design method for single-phase photovoltaic systems connected to the grid via micro-inverters. Technologies 12(6):89
52. Gellings CW (2020) The smart grid: enabling energy efficiency and demand response. River Publishers
53. Sutikno T, Arsadiando W, Wangsupphaphol A, Yudhana A, Facta M (2022). A review of recent advances on hybrid energy storage system for solar photovoltaics power generation. IEEE Access 10:42346–42364
54. Hamzat AK, Sahin AZ, Omisanya MI, Alhems LM (2021) Advances in PV and PVT cooling technologies: a review. Sustain Energy Technol Assess 47:101360
55. Arbouch I, Karzazi Y, Hammouti B (2014) Organic photovoltaic cells: operating principles, recent developments and current challenges–review. Phys Chem News 72(4):73–84
56. Kim JY, Lee JW, Jung HS, Shin H, Park NG (2020) High-efficiency perovskite solar cells. Chem Rev 120(15):7867–7918
57. Al-Ezzi AS, Ansari MNM (2022) Photovoltaic solar cells: a review. Appl Syst Innov 5(4):67
58. Blakers A, Zin N, McIntosh K, Fong K (2013) High efficiency silicon solar cells. Energy Procedia 33:1–10
59. Kolamroudi MK, Ilkan M, Egelioglu F, Safaei B (2023) Feature selection by ant colony optimization and experimental assessment analysis of PV panel by reflection of mirrors perpendicularly. Renew Energy 218:119238
60. Menke SM, Holmes RJ (2014) Exciton diffusion in organic photovoltaic cells. Energy Environ Sci 7:499–512
61. Sharaf M, Yousef MS, Huzayyin AS (2022) Review of cooling techniques used to enhance the efficiency of photovoltaic power systems. Environ Sci Pollut Res 29(18):26131–26159
62. Ko J, Kim K, Sohn JW, Jang H, Lee HS, Kim D, Kang Y (2023) Review on separation processes of end-of-life silicon photovoltaic modules. Energies 16(11):4327
63. Teh CJQ, Drieberg M, Soeung S, Ahmad R (2021) Simple PV modeling under variable operating conditions. IEEE Access 9:96546–96558
64. Sivaraj S, Rathanasamy R, Kaliyannan GV, Panchal H, Jawad Alrubaie A, Musa Jaber M, Said Z, Memon S (2022) A comprehensive review on current performance, challenges and progress in thin-film solar cells. Energies 15(22):8688
65. Waller R, Kacira M, Magadley E, Teitel M, Yehia I (2022) Evaluating the performance of flexible, semi-transparent large-area organic photovoltaic arrays deployed on a greenhouse. AgriEngineering 4(4):969–992
66. Hirst LC, Ekins-Daukes NJ (2011) Fundamental losses in solar cells. Progr Photovolt: Res Appl 19(3):286–293
67. Wen L, Li G, Huang T, Geng W, Pei H, Yang J, Zhu M, Zhang P, Hou R, Tian G, Su W (2022) Single-cell technologies: from research to application. The Innovation 3(6)
68. Rahman MM, Oni AO, Gemechu E, Kumar A (2020) Assessment of energy storage technologies: a review. Energy Convers Manag 223:113295
69. Akhtar I, Asim M, Yadav RK, Agarwal P, Kirmani S (2019, August) Design of effective grid-connected solar system. In: International conference on inventive computation technologies. Springer International Publishing, Cham, pp 606–614
70. Feron S (2016) Sustainability of off-grid photovoltaic systems for rural electrification in developing countries: a review. Sustainability 8(12):1326
71. Ciocia A, Amato A, Di Leo P, Fichera S, Malgaroli G, Spertino F, Tzanova S (2021) Self-consumption and self-sufficiency in photovoltaic systems: effect of grid limitation and storage installation. Energies 14(6):1591
72. Litjens GB, Worrell E, Van Sark WGJHM (2018) Economic benefits of combining self-consumption enhancement with frequency restoration reserves provision by photovoltaic-battery systems. Appl Energy 223:172–187

73. Vieira FM, Moura PS, de Almeida AT (2017) Energy storage system for self-consumption of photovoltaic energy in residential zero energy buildings. Renew Energy 103:308–320
74. Bakelli Y, Arab AH, Azoui B (2011) Optimal sizing of photovoltaic pumping system with water tank storage using LPSP concept. Solar Energy 85(2):288–294
75. Karthikeyan V, Rajasekar S, Das V, Karuppanan P, Singh AK (2017) Grid-connected and off-grid solar photovoltaic system. In: Smart energy grid design for island countries: challenges and opportunities. Springer International Publishing, Cham, pp 125–157
76. Wichert B, Dymond M, Lawrance W, Friese T (2001) Development of a test facility for photovoltaic-diesel hybrid energy systems. Renew Energy 22(1–3):311–319
77. Ridha HM, Gomes C, Hizam H, Ahmadipour M, Muhsen DH, Ethaib S (2020) Optimum design of a standalone solar photovoltaic system based on novel integration of iterative-PESA-II and AHP-VIKOR methods. Processes 8(3):367
78. Florida Solar Energy Center (n.d.) Types of PV systems. Retrieved from https://energyresearch.ucf.edu
79. Khlifi F, Cherif H, Belhadj J (2021) Environmental and economic optimization and sizing of a micro-grid with battery storage for an industrial application. Energies 14(18):5913
80. Zarina PP, Mishra S, Sekhar PC (2014) Exploring frequency control capability of a PV system in a hybrid PV-rotating machine-without storage system. Int J Electr Power Energy Syst 60:258–267
81. Jeong KS, Lee WY, Kim CS (2005) Energy management strategies of a fuel cell/battery hybrid system using fuzzy logics. J Power Sources 145(2):319–326
82. Hredzak B, Agelidis VG (2012) Direct current control of a battery-ultracapacitor power supply. In: IECON 2012-38th annual conference on IEEE industrial electronics society. IEEE, pp 4024–4028
83. Lahyani A, Venet P, Guermazi A, Troudi A (2012) Battery/supercapacitors combination in uninterruptible power supply (UPS). IEEE Trans Power Electron 28(4):1509–1522
84. Kadri A, Marzougui H, Aouiti A, Bacha F (2020) Energy management and control strategy for a DFIG wind turbine/fuel cell hybrid system with super capacitor storage system. Energy 192:116518
85. Ortúzar M, Moreno J, Dixon J (2007) Ultracapacitor-based auxiliary energy system for an electric vehicle: implementation and evaluation. IEEE Trans Ind Electron 54(4):2147–2156
86. Teleke S, Baran ME, Bhattacharya S, Huang AQ (2010) Rule-based control of battery energy storage for dispatching intermittent renewable sources. IEEE Trans Sustain Energy 1(3):117–124
87. Chong LW, Wong YW, Rajkumar RK, Rajkumar RK, Isa D (2016) Hybrid energy storage systems and control strategies for stand-alone renewable energy power systems. Renew Sustain Energy Rev 66:174–189
88. Jing W, Lai CH, Wong WS, Wong MD (2018) A comprehensive study of battery-supercapacitor hybrid energy storage system for standalone PV power system in rural electrification. Appl Energy 224:340–356
89. Asensio EM, Magallán GA, De Angelo CH, Serra FM (2020) Energy management on battery/ultracapacitor hybrid energy storage system based on adjustable bandwidth filter and sliding-mode control. J Energy Storage 30:101569
90. Jing W, Hung Lai C, Wong SHW, Wong MLD (2017) Battery-supercapacitor hybrid energy storage system in standalone DC microgrids: a review. IET Renew Power Gener 11(4):461–469
91. Xie P, Tan S, Guerrero JM, Vasquez JC (2021) MPC-informed ECMS based real-time power management strategy for hybrid electric ship. Energy Rep 7:126–133
92. Muqeet HA, Javed H, Akhter MN, Shahzad M, Munir HM, Nadeem MU, Bukhari SSH, Huba M (2022) Sustainable solutions for advanced energy management system of campus microgrids: model opportunities and future challenges. Sensors 22(6):2345
93. Zhang J, Huang L, Shu J, Wang H, Ding J (2017) Energy management of PV-diesel-battery hybrid power system for island stand-alone micro-grid. Energy Procedia 105:2201–2206

94. Aziz AS, Tajuddin MFN, Adzman MR, Ramli MA, Mekhilef S (2019) Energy management and optimization of a PV/diesel/battery hybrid energy system using a combined dispatch strategy. Sustainability 11(3):683
95. Emara D, Ezzat M, Abdelaziz AY, Mahmoud K, Lehtonen M, Darwish MM (2021) Novel control strategy for enhancing microgrid operation connected to photovoltaic generation and energy storage systems. Electronics 10(11):1261
96. Wang Y, Jiao X (2022) Dual heuristic dynamic programming based energy management control for hybrid electric vehicles. Energies 15(9):3235
97. Dai K, Zhang C, Li Q, Huang X, Zhang H (2021) An adaptive energy management strategy for simultaneous long life and high wake-up success rate of wireless sensor network nodes. Energy Technol 9(10):2100522
98. Gharibeh HF, Yazdankhah AS, Azizian MR, Farrokhifar M (2021) Online energy management strategy for fuel cell hybrid electric vehicles with installed PV on roof. IEEE Trans Ind Appl 57(3):2859–2869
99. Ahmed S, Muhammad Adil HM, Ahmad I, Azeem MK, Huma Z, Abbas Khan S (2020) Supertwisting sliding mode algorithm based nonlinear MPPT control for a solar PV system with artificial neural networks based reference generation. Energies 13(14):3695
100. Gaviria JF, Narváez G, Guillen C, Giraldo LF, Bressan M (2022) Machine learning in photovoltaic systems: a review. Renew Energy 196:298–318
101. Berghout T, Benbouzid M, Bentrcia T, Ma X, Djurović S, Mouss LH (2021) Machine learning-based condition monitoring for PV systems: state of the art and future prospects. Energies 14(19):6316
102. Garoudja E, Chouder A, Kara K, Silvestre S (2017) An enhanced machine learning based approach for failures detection and diagnosis of PV systems. Energy Convers Manag 151:496–513
103. Tina GM, Ventura C, Ferlito S, De Vito S (2021) A state-of-art-review on machine-learning based methods for PV. Appl Sci 11(16):7550
104. Moyo RT, Tabakov PY, Moyo S (2021) Design and modeling of the ANFIS-based MPPT controller for a solar photovoltaic system. J Solar Energy Eng 143(4):041002
105. Farah L, Haddouche A, Haddouche A (2020) Comparison between proposed fuzzy logic and ANFIS for MPPT control for photovoltaic system. Int J Power Electron Drive Syst 11(2):1065
106. Ramadan A, Kamel S, Hamdan I, Agwa AM (2022) A novel intelligent ANFIS for the dynamic model of photovoltaic systems. Mathematics 10(8):1286
107. Kim S, Chou PH (2015) Energy harvesting with supercapacitor-based energy storage. In: Smart sensors and systems. Springer International Publishing, Cham, pp 215–241
108. Sinha S, Bajpai P (2020) Power management of hybrid energy storage system in a standalone DC microgrid. J Energy Storage 30:101523
109. Hernández JC, Gomez-Gonzalez M, Sanchez-Sutil F, Jurado F (2021) Optimization of battery/supercapacitor-based photovoltaic household-prosumers providing self-consumption and frequency containment reserve as influenced by temporal data granularity. J Energy Storage 36:102366
110. Kollimalla SK, Mishra MK, Narasamma NL (2014) Design and analysis of novel control strategy for battery and supercapacitor storage system. IEEE Trans Sustain Energy 5(4):1137–1144
111. Guo XQ, Wu WY, Gu HR (2011) Phase locked loop and synchronization methods for grid-interfaced converters: a review. Przegląd Elektrotechniczny 87:182–187
112. Chenni R, Makhlouf M, Kerbache T, Bouzid A (2007) A detailed modeling method for photovoltaic cells. Energy 32(9):1724–1730
113. Gergaud O (2002) Modélisation énergétique et optimisation économique d'un système de production éolien et photovoltaïque couplé au réseau et associé à un accumulateur. Doctoral dissertation, École normale supérieure de Cachan-ENS Cachan
114. Doerffel D, Sharkh SA (2006) A critical review of using the Peukert equation for determining the remaining capacity of lead-acid and lithium-ion batteries. J Power Sources 155(2):395–400

115. Tremblay O, Dessaint LA, Dekkiche AI (2007) A generic battery model for the dynamic simulation of hybrid electric vehicles. In: 2007 IEEE vehicle power and propulsion conference, vol 284. IEEE
116. Widanage WD, Barai A, Chouchelamane GH, Uddin K, McGordon A, Marco J, Jennings P (2016) Design and use of multisine signals for Li-ion battery equivalent circuit modelling. Part 2: Model estimation. J Power Sources 324:61–69
117. Luo Z, Li Y, Lou Y (2015) An adaptive Kalman filter to estimate state-of-charge of lithium-ion batteries. In: 2015 IEEE international conference on information and automation. IEEE, pp 1227–1232
118. Hussein AAH, Batarseh I (2011) An overview of generic battery models. In: 2011 IEEE power and energy society general meeting. IEEE, pp 1–6
119. Hadjipaschalis I, Poullikkas A, Efthimiou V (2009) Overview of current and future energy storage technologies for electric power applications. Renew Sustain Energy Rev 13(6–7):1513–1522
120. Cabrane Z, Ouassaid M, Maaroufi M (2016) Analysis and evaluation of battery-supercapacitor hybrid energy storage system for photovoltaic installation. Int J Hydrog Energy 41(45):20897–20907
121. Lahyani A, Sari A, Lahbib I, Venet P (2016) Optimal hybridization and amortized cost study of battery/supercapacitors system under pulsed loads. J Energy Storage 6:222–231
122. Negroiu R, Svasta P, Vasile A, Ionescu C, Marghescu C (2016) Comparison between Zubieta model of supercapacitors and their real behavior. In: 2016 IEEE 22nd international symposium for design and technology in electronic packaging (SIITME). IEEE, pp 196–199
123. Dhumale RB, Thombare ND, Lokhande SD (2017) Modeling and diagnosis of open switch fault in three phase VSI. In: 2017 2nd IEEE international conference on recent trends in electronics, information & communication technology (RTEICT). IEEE, pp 314–319
124. Diagne S, Farota AK, Assogba O, Diop B, Val T (2021) Solar energy recovery and storage system for powering wireless communicating nodes. Energy Power Eng 13(12):377–391
125. Yu N, Yang J, Chen S, Ye M (2014) Design of LCL-filter based three-level active power filter. TELKOMNIKA Indonesian J Electr Eng 12(1):48–56
126. Li Y, Zhang J, Hao Z, Tian P (2021) Improved PR control strategy for an LCL three-phase grid-connected inverter based on active damping. Appl Sci 11(7):3170
127. Ghasemi MA, Zarei SF, Peyghami S, Blaabjerg F (2021) A theoretical concept of decoupled current control scheme for grid-connected inverter with LCL filter. Appl Sci 11(14):6256
128. Zdiri MA, Guesmi T, Alshammari BM, Alqunun K, Almalaq A, Salem FB, Abdallah HH, Toumi A (2022) Design and analysis of sliding-mode artificial neural network control strategy for hybrid PV-battery-supercapacitor system. Energies 15(11):4099
129. Young KD, Utkin VI, Ozguner U (1999) A control engineer's guide to sliding mode control. IEEE Trans Control Syst Technol 7(3):328–342
130. Ciaburro G (2024) MATLAB for Machine Learning: Unlock the power of deep learning for swift and enhanced results. Packt Publishing Ltd
131. Standards Coordinating Committee 21 (2003) 1547TM IEEE standard for interconnecting distributed resources with electric power systems. The Institute of Electrical and Electronics Engineers. https://fglongatt.org/OLD/Archivos/Archivos/SistGD/IEEE1547.pdf
132. Priemer R (2013) MATLAB for electrical and computer engineering students and professionals with Simulink. The Institution of Engineering and Technology. https://doi.org/10.1049/SBPC501E

Data-Driven Condition Monitoring of Reciprocating Compressors Using Snapshot Techniques and Fault Classification

Al-Haitham Al-Obaidani, Morteza Mohammadzaheri, and Ali Al-Humairi

Abstract Reciprocating compressors are important machinery used in the oil and gas industry but are prone to wear and failure due to their many moving parts. Traditional maintenance methods like reactive and scheduled inspections often miss early signs of faults, which can lead to unplanned downtime, costly repairs, and safety risks. Predictive maintenance offers a more proactive approach by identifying potential faults before failure can happen. This study explores a data-driven approach for fault detection using snapshot data from an open-source dataset of a two-cylinder, single-stage reciprocating air compressor. The dataset contains sensor data and health status labels for four components: bearings, radiator, water pump, and discharge valve. A multi-layer perceptron (MLP) model was built to classify the condition of each component as either healthy or unhealthy. The architecture consisted of 20 input neurons, two hidden layers, and 4 neurons in the output layer which corresponds to each component. The model was tested using accuracy, loss, precision, and recall metric for evaluation. Bayesian optimization was applied to reduce the size of the parameters for a more efficient performance. The final model achieved an average macro $F1$-Score of 0.991 and a 1.0 recall for all the components. These results show that snapshot-based fault classification has potential to assist operators with early fault detections and better maintenance decisions.

Keywords Reciprocating compressors · Predictive maintenance · Fault classification · Condition monitoring · Snapshot monitoring · Machine learning · Artificial neural network (ANN) · Data-driven modeling · Multi-layer perceptron (MLP)

Abbreviations

MLP	Multi-layer perceptron

A.-H. Al-Obaidani · M. Mohammadzaheri (✉) · A. Al-Humairi
German University of Technology in Oman, Halban, Sultanate of Oman
e-mail: morteza.zaheri@gutech.edu.om

A. Azizi (ed.), *Smart Diagnostics and Predictive Maintenance*, Emerging Trends in Mechatronics, https://doi.org/10.1007/978-981-95-3921-5_2

BCE	Binary cross-entropy
HL	Hidden layer
RPM	Revolutions per minute
AC	Alternating current
$F1$-Score	F1 performance metric
ReLU	Rectified linear unit
PV	Pressure–volume
OEM	Original equipment manufacturer
FFT	Fast Fourier transform
RF	Random forest
SVM	Support vector machine
CNN	Convolutional neural network
ML	Machine learning
DL	Deep learning
PCA	Principle component analysis
ANN	Artificial neural network
LSTM	Long-term short memory
LMD	Local mean decomposition
KNN	K-nearest neighbors

Symbols

X	Original value of a feature
$X_{\min}$	Minimum value in the feature column
$X_{\max}$	Minimum value in the feature column
X_n	Normalized value of the feature
y_i	Actual binary label for the ith sample
P_i	Predicted probability for the ith sample
TP	True Positive (Model predicts faulty correctly)
FP	False positive (Model predicts faulty when healthy)
FN	False negative (Model misses faulty condition)
$f(x)$	Rectified linear activation function
N	Number of samples

1 Introduction

1.1 Overview

This section provides a detailed review of the background information and related studies needed to understand the challenges and solutions related to condition monitoring of reciprocating compressors. To understand the concept of condition monitoring, the section begins by explaining what reciprocating compressors are, how they work, and what are the main components that are involved in the compression operation.

After that, the section discusses the most common failure modes seen in reciprocating compressors, which includes insights on which components cost more to maintain, and which causes most compressor shutdowns. The section then moves on to introduce the different types of maintenance strategies that are used in the industry and explains how predictive maintenance and condition monitoring can be used for fault diagnosis with the help of data collection using sensors.

In this part, several monitoring techniques are covered which include vibration, ultrasonic, pressure, and temperature. This is accompanied by past research applications that used these techniques to detect failure in reciprocating compressors.

The section ends by reviewing existing research on the use of condition monitoring techniques and discussing the application of machine learning for fault classification.

1.2 Overview of Reciprocating Compressors

1.2.1 Type

Reciprocating compressors can be divided into single-stage and multi-stage types based on how many steps it takes to reach the final pressure [4]. According to Bloch and Hoefner [5], in a single-stage compressor, the gas is compressed from the starting pressure to the final pressure in just one step. However, in a multi-stage compressor, the gas passes through two or more compression stages till it reaches to the final pressure.

As the gasses are compressed to higher pressures, their temperature also increases, which creates operational problems related to discharge pressure, allowable pressure rise, and compression ratios. Bloch and Hoefner [5] state that using a multi-stage compressor is a better option for these situations because it saves power and reduces temperature and differential pressures. Nwakpang et al. [17] adds to this by saying that compression process is more efficient and also saves energy.

Bloch and Hoefner [5] goes on to say that the compression process in a reciprocating compressor can take place on either one side of the piston (single-acting) or both sides (double-acting). Single-acting compressors are used in moderate-duty applications, where the compressor is usually air-cooled and not continuously running at full capacity. On the other hand, double-acting compressors are used in heavy-duty applications, where they operate at high loads for extended periods. These types should be water-cooled to manage the continuous heat generation during the operation.

1.2.2 Working Principle

Stewart [28] explains that pressure inside the cylinder changes with volume during one full crankshaft revolution, showing the different stages of the cycle which include suction, compression, discharge, and expansion. This relationship can be presented using a P–V diagram that shows the complete compression cycle as shown in Fig. 1.

The cycle starts when the piston is at the bottom dead center, which is the lower end of the stroke, and the cylinder is filled with gas at the suction pressure. The piston then starts moving and compressing the gas, increasing the pressure from P_1 to P_2 (Point 1 to 2). During this process, both the suction and the discharge valves are closed.

Once the piston reaches the top dead center (Point 2), which is the upper end of the stroke, the pressure inside the cylinder is now more than the pressure in the discharge line, which cause the valves to open automatically, allowing the gas to escape from inside the cylinder at the discharge pressure P_2 (Point 2 to 3).

After the discharge process, there is a small amount of gas remaining inside the cylinder between the piston and the head end of the cylinder, that space is called the clearance volume. As the piston starts moving back, that remaining gas expands, and pressure inside the cylinder decreases from P_2 to P_1 (Point 3 to 4). Similar to the compression stage, both the valves are closed in the expansion stage.

When the piston reaches the end of the expansion stroke, the pressure inside the cylinder becomes lower than the pressure in the suction line causing the suction valve to open and gas starts flowing in (Point 4 to 1). When the cylinder is filled again, the cycle is repeated with the next revolution [22, 28].

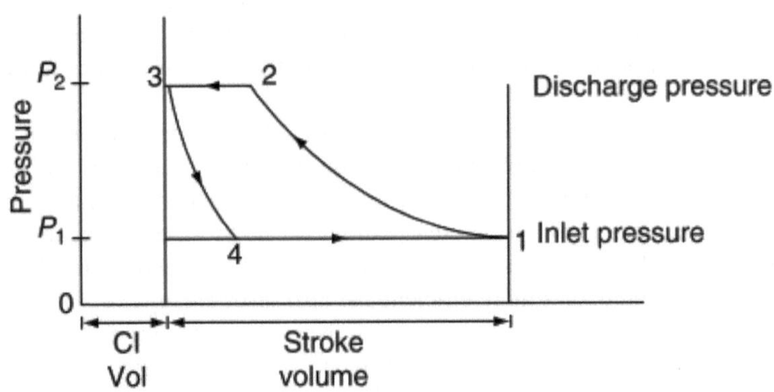

Fig. 1 P–V compression cycle [28]

1.2.3 Major Components

The major components of reciprocating compressors are the crankcase, crankshaft, bearings, crosshead, piston, piston rod, connecting rod, cylinder, and valves [5].

The crankcase acts as the structural base if the reciprocating compressor. It is made from a high-grade iron cast that holds and supports all the moving and rotating parts such as the crankshaft, crosshead, and connecting rod. The frame has internal supports that hold these components to keep them properly aligned to manage the mechanical stresses that occur during the operation [5, 22, 28].

The cylinder in a reciprocating compressor is a pressure vessel that contains the gas during the compression cycle, and it is where the piston moves back and forth to compress the gas [28]. The design of the cylinder takes into consideration the operating pressure, bore size, and the type of gas that is used [5]. The material used for the cylinder depends on the application in which the compressor is used for. Bloch and Hoefner [5] mentions that cast iron is used for low-pressure cylinders, while steel forging is used for high-pressure cylinders for pressure ranges over 2000 psi. Royce [22] adds to this by suggesting that nodular or ductile iron can also be used instead of cast iron.

Inside the cylinder, there is usually a liner that acts as the wear surface which the piston operates against. The liner is designed to provide a renewable surface that extends the life of the cylinder, and it can be replaced. This allows maintenance operators to replace only the liner when wear occurs, rather than replacing the whole cylinder. This component is most commonly adopted in larger or multi-stage compressors [22]. Cylinders are also equipped with cooling systems to manage the heat generated during compression. Water jackets are used for in most compressors, while air cooling is used for smaller systems [22].

Inside the cylinder, there is a component which is known as the liner. The liner simply acts as the surface in which the piston moves back and forth against. Instead of the piston causing wear on the cylinder itself, the liner is basically a protective layer that takes the wear and can be replaced, extending the life of the cylinder. They are mostly installed in larger or multi-stage compressors [22].

Cylinders also contain cooling systems that are responsible for managing heat generated during the compression cycle. Most compressors use water jackets for cooling, while air cooling is mainly used for smaller systems [22].

The crankshaft is the part that turn rotational motion from the driver into the reciprocating motion needed to move the piston. It is usually made from either forged carbon steel or nodular iron casting to handle the stresses of the operation [5, 28]. Royce [22] mentions that crankshafts should be equipped with removable counterweights to balance out the rotary and reciprocating forces, so the compressor is able to run smoothly. The connecting rod simply connects the crankshaft to the crosshead, while crosshead converts the rotational motion of the connecting rod to linear motion of both the piston rod and piston. The piston rod pushes the piston back and forth within the compressor cylinder [28].

The piston plays one of the most important roles of the compressor as they must transfer the energy from the crankshaft to the gas, compressing it within the cylinder

[22]. According to Stewart [28], the material used to make the piston depend on its design and application. Lightweight materials like aluminum, cast iron, or steel are used. Bloch and Hoefner [5] adds to this by stating that aluminum is chosen when reducing weight is the main concern to help balance the reciprocating motion. Cast iron is used for the strength and wear resistance it provides. Steel is typically used for small-bore, where higher strength is needed.

To make sure there is no leakage in the process, the piston is fitted with rings that create a seal between the piston and cylinder walls which also minimizes the wear by allowing the piston to smoothly move back and forth [22]. Rider rings are also installed to support the weight of the piston, help guide the piston within the cylinder and prevent direct contact of the piston and the cylinder wall. Both piston and rider rings should be made out of the same material which are usually made from Teflon [28].

The valves control the flow of gas in and out of the cylinder. Suction valves allow gas into the cylinder before compression, while discharge valves allow gas out of the cylinder after compression. This flow is restricted only to one direction meaning that the gas is not allowed in from the discharge valves or out from the suction valve [28]. Royce [22] explains that valves need to be made from durable materials that last long operations. Since compressors usually deal with different types of gasses, the material should also be compatible with the type of gas used.

Inside the valves are multiple components that control the flow in and out of the cylinder. The valve seats are the part that seals the valve closed so that the gas does not leak backwards. They must be strong enough to handle the pressure differentials between the inner and outer parts of the valve and the wear from the constant opening and closing. Valve plates move back and forth to let gas flow in and out through the valve. Valve spring help close the valves when the pressure differentials approach zero. Their design must suit the specific conditions within the cylinder to avoid any safety issues. The guard guides the motion and handle the impact forces of the valve plate [5].

1.2.4 Typical Failure Modes

Reciprocating compressors can experience different types of failures. As shown in Fig. 2, typical failure modes include valves, packings, and ring leakage; foundation and bearing problems, and pulsations.

According to Stewart [28], even though reciprocating compressors only make up less than 5% of the equipment in most plants, they are the cause for more than 15% of maintenance budget and most of that goes toward valves.

[27] provide a breakdown of maintenance cost percentages for each major component in the reciprocating compressor, showing which parts contribute the most toward the maintenance budget. These results are shown in Table 1.

Table 2 shows a survey conducted by [12], where over 200 users gave their experience with reciprocating compressor components that caused the most shutdowns.

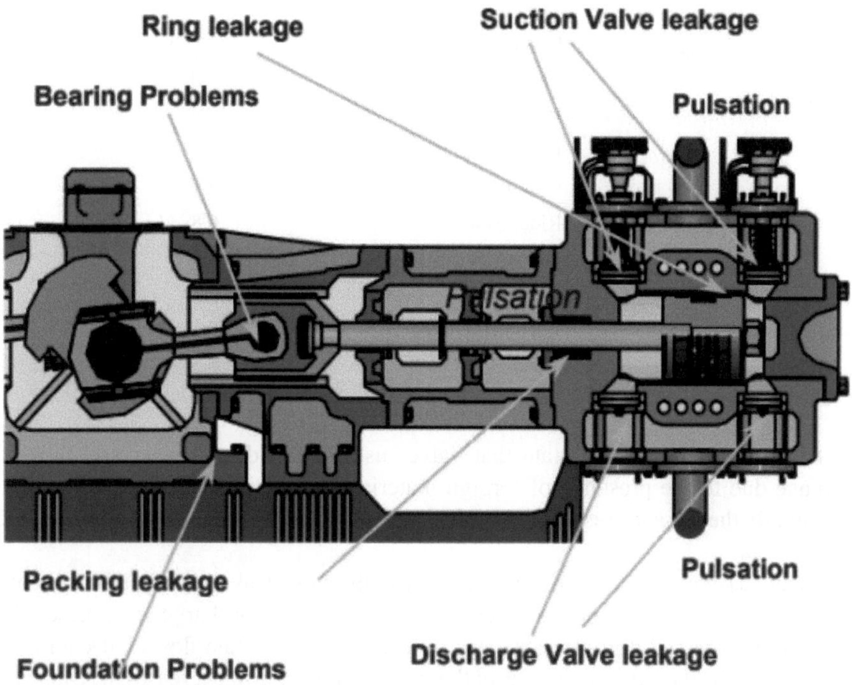

Fig. 2 Typical failure modes (courtesy of HOERBIGER)

Table 1 Maintenance costs per component [27]

Component	Cost percent (%)
Valves	50
Packing	20
Piston rings	20
Rider rings	7
Piston rods	2
Liners	0.5
Bearings	0.5

The table clearly shows that valve-related issues contribute the most to unexpected shutdowns when compared to the other components.

Looking at the cited literature, it is clear the valves account for the largest share of the maintenance budget and are responsible for most shutdowns compared to the other components. [5] highlights that valves have the greatest effect on both the efficiency and mechanical reliability of a reciprocating compressor. Valves must open and close with every stroke of the piston, often a thousand times per hour, hence, they are subjected to all kinds of destructive forces.

Table 2 Components causing the most shutdowns [12]

Component	Percent (%)
Valves	36.0
Packings	17.8
Process problems	8.8
Piston rings	7.1
Rider band	6.8
Unloaders	6.8
Cylinder lube systems	5.1
Instrumentations	5.1
Other	6.5

Bloch and Hoefner [5] state that valves usually fail due to wear and fatigue, damage due to the presence of foreign materials, or abnormal mechanical action. Eventually these issues result in valve leakage which impacts the performance of the compressor.

Bloch [4] explains in detail how suction and discharge leakages affect the capacity and efficiency of the reciprocating compressor. When the discharge valve leaks, the already compressed high-pressure gas from the discharge line flows back into the cylinder during the expansion stage of the cycle. This backflow decreases the capacity. Likewise, when the suction valve leaks the gas leaves the cylinder to the suction line which also reduces the capacity. Schirmer [25] mentions that valve leakage can easily be identified when comparing the actual P–V diagram with the ideal P–V diagram. They add to this by claiming that valve leakage cannot be avoided, and the amount of gas going in the cylinder is almost never equal to the amount of gas leaving. Bloch [4] argues that a zero-leakage valve design is a worthwhile goal to pursue because even with minimum leakages the impact can be significant on the reciprocating compressor performance.

The second highest contributors to both maintenance costs and shutdowns are the piston packing rings. Packings are used to seal the space around the piston rod as it moves in and out of the cylinder. Because of this sliding motion of the piston rod, there is friction between the rod and the inner surface of the packing rings which cause wear. This wear eventually causes leakage and similar to valve leakage this results in lower capacity [4]. Leakage can also be a result of improper assembly and if wear occurs on the piston rod [5].

Leakage can also occur in piston rings. When piston rings wear or break, gas can escape past the piston into the other side of the cylinder which lowers the pressure in the compression side. This pressure drop lowers the capacity and increases the gas temperature within the cylinder because this high-temperature gas leaked heats the incoming gas from the expansion stage [4].

1.3 Predictive Maintenance Approaches

1.3.1 Types of Maintenance

According to Mobley [15], maintenance management can be split into three main methods: run to failure (reactive), preventative maintenance, and predictive maintenance. In reactive maintenance, the compressor is allowed to operate until something fails. Any minor problems are ignored, and maintenance is performed only after a failure has occurred. While this method offers maximum usage for maximum production output from the compressor, it is expensive, inefficient, less safe, and has longer downtime when compared to other types of maintenance [31]. Mobley [15] also states that reactive maintenance costs are three times greater when compared to scheduled maintenances.

Preventative maintenance is time-based and is carried out at regular time intervals [15, 31], and it does not matter what the current condition of the compressor is. These maintenance intervals are either based on the Original Equipment Manufacturer (OEM) recommendations or experience from past operations. This method minimizes the risk of sudden failure, which in return reduces downtime and costs but it can also lead to unnecessary repairs [31].

Predictive maintenance is a condition-based preventative maintenance approach [15] that uses operational data that is collected using different sensors that are mounted on various components on the compressor [31]. Using this data, different monitoring tools are used to determine the condition of the compressor [26]. Measurements like pressure within the cylinder, vibration, temperature, and ultrasonic are monitored and used to detect early signs of failure [6, 26, 31]. By comparing these data with data from normal operation, the operator can know the exact condition of the compressor and can schedule maintenance just before potential failure occurs and perform the necessary repairs. Although using this approach minimizes downtime, high maintenance costs, and unnecessary maintenance, the cost of the monitoring devices is high [31].

1.3.2 Condition-Based Monitoring Techniques

The main concept of condition-based monitoring involves using sensors to measure physical variables such as temperature, vibration, and pressure from an operating machine [19, 25]. This allows operators to compare the data collected with baseline references or historical trends to determine the condition of the current equipment and detect potential faults before they can occur.

Griffith and Flanagan [8], Schultheis et al. [26], and Bloch [4] cover several techniques that are used to determine the condition of reciprocating compressors, and these include: vibration, ultrasonic, pressure, and temperature monitoring.

Vibration monitoring involves using accelerometers to collect vibration data, which can be analyzed either in the time domain or frequency domain [2]. For reciprocating compressors, time domain analysis is referenced to the position of the crankshaft and shows how vibration patterns vary throughout the compression cycle [8]. While frequency domain analysis helps identify specific fault frequencies using Fast Fourier Transform (FFT) analysis [4, 15].

Accelerometers are usually mounted on crosshead housing, crankcase [8, 26], cylinder, and valve covers [4]. Vibration measurements from the crosshead housing help identify issues such as loose pistons, rider wear, ring wear, and crosshead problems [26], and sometimes cylinder issues for small to medium-sized compressors [21]. As for the accelerometer mounted on the crankcase, they help identify foundation issues, frame misalignments and mechanical problems [8]. The accelerometers mounted on the valve covers help detect faults such as mechanical damage, delayed opening and closing of the valves, spring failure, and plate impact [11, 15].

Ultrasonic monitoring is an extension vibration monitoring and is mainly used to evaluate valve condition in early stages before a failure occurs [26]. This is because ultrasonic microphones detect high-frequency acoustic signals that can be generated by turbulent flow of gas which are caused by leakage [15]. When mounted on valve covers, leakage in the valve can be identified. On cylinder ends, leaks from gaskets, rings, and packings can be detected [15]. Like vibration monitoring, the ultrasonic measurements are also synced with crankshaft position [8]. In addition, when these measurements are combined with pressure patterns, operators can know how these leaks impact compression efficiency and energy losses [26].

Pressure monitoring involves collecting pressure measurements using pressure transducers mounted on the head and crank ends of the cylinder allows operators to assess the performance of valves, piston rings, and packings [26]. Schirmer [25] states in their paper that pressure against volume is the best way to show the energy cycle in a reciprocating machine. PV diagrams help operators calculate the amount of power required for the compression cycle, capacity, and volumetric efficiencies [25]. In addition, Griffith and Flanagan [8] mentioned in their paper that by comparing a PV diagram of an operating compressor with a theoretically calculated PV diagram, faults can be diagnosed.

Temperature monitoring involves collecting temperatures at the cylinder, valves, and packings. Infrared thermometers mounted on the valve covers and suction and discharge flanges and are used to verify leakage diagnosis from ultrasonic monitoring. In addition, taking temperature at packing casing can identify leakage in packing [26]. Additionally, these measurements also help assess the thermodynamic performance of the compressor [25].

The data for these techniques are collected either periodically or continuously [25], it actually depends on the design of the monitoring system and the criticality of the equipment used [26]. For example, a main reciprocating compressor that supplies natural gas into a refinery pipeline is considered high criticality. If it fails, it stops the supply of gas which disrupts the whole process. In this case, the compressor is equipped with an online (continuous) monitoring system to track the condition of the compressor in real time.

On the other hand, a reciprocating compressor used to supply air to pneumatic tools in a workshop is considered low criticality. If it fails, it can cause a delay in tool operation and the disruption is less when compared to the previous example. For this, a periodic approach works.

1.3.3 Applications of Condition Monitoring Techniques in Fault Diagnosis

Many research studies have demonstrated how these sensor data can be used to monitor compressor health and also detect and classify failure modes. This section lists a few examples from existing literature that show how these techniques were used to identify and differentiate between healthy and faulty operating conditions.

Guerra [9] conducts an experiment on reciprocating compressor valves by introducing seeded faults like valve wear and spring fatigue. These faults are then monitored using different sensors mounted around the valve to monitor the vibration, temperature, and pressure. Out of all the sensors, the dynamic pressure sensor performed the best in differentiating the health conditions of the valve with a classification of 93.2% for the two different failure modes. Guerra also noted that the accelerometer showed potential in showing the different frequency peaks for the different valve health states, but the machine was too noisy which interfered with the classification.

Qi et al. [20] collected 5 years' worth of operational data such as vibration, pressure, temperature, noise amplitude, and rod position to detect valve failures and piston rod issues in reciprocating compressors. Using machine learning techniques, their system achieved over 80% classification accuracy.

Ahmed et al. [1] conducted vibration monitoring using accelerometers to detect valve-related issues. They processed the data using statistical analysis and were able to identify all the fault types correctly.

1.4 Application of Machine Learning in Fault Diagnosis

Traditionally, fault diagnosis in reciprocating compressors relied on human experts that manually interpret anomalies in the compressor and comparing them to the baseline performance [14]. A common traditional fault diagnosis method is rule-based systems like fuzzy logic. Bisal and Jana [3] proposed an approach that combines both fuzzy logic and random forest (RF) to monitor the health of the compressor and predict maintenance schedules. Their system was able to predict when the upcoming maintenance is needed and grade the severity of anomalies based on criticality levels.

Another common traditional method is using statistical analysis. Ahmed et al. [1] developed a fault detection method using principal component analysis (PCA) on vibration data. They measured the vibration signals of normal operation and five different seeded fault conditions and use PCA to reduce the complexity of the dataset. Statistical tests (T^2 and Q) were applied to detect how each fault deviates from normal operation and were able to identify each fault condition.

Kolodziej and Trout [11] applied a traditional ML technique called the Bayesian classifier to predict the severity of valve seat wear failure. They use Short-time Fourier transform to turn raw vibrations into heatmap pictures that are then analyzed to extract the features of the image for classification. Their model has a success rate of 98%.

Another traditional ML technique is support vector machine (SVM). Zhao et al. [30] use SVM for fault diagnosis on four different bearing clearance states by using local mean decomposition (LMD) to extract different features which were then used to train the SVM model. One model achieved 95% total accuracy, and the other 3 were all around 91%.

Leoni et al. [13] aimed to enhance fault diagnoses in reciprocating compressors by eliminating noise and implementing data reduction techniques before using RF classifier to identify fault conditions. This is tested on a compressor which is subjected to surge, and the accuracy achieved on both training and testing sets was around 97.3%.

As observed, all the traditional machine learning techniques reviewed required manual feature extraction process. On the other hand, modern ML techniques like deep learning models can automatically learn and extract features directly from the raw data [7].

Sánchez et al. [23] compare traditional ML with a one-dimensional convolutional neural networks (CNN) which is a deep learning (DL) model. They use a combination of feature extractions and two different ML classifiers that include RF and K-nearest neighbors (KNN). Both models are trained to classify ten faults of valve and bearing issues. Results showed that DL performed way better with an average classification accuracy of 90.7% opposed to 86% that is achieved by KNN and 85% achieved by RF.

Saturday and Callistus [24] built an artificial neural network (ANN) to predict failure of a reciprocating compressor using three months of operational data. The architecture of the model consisted of an input layer with 10 neurons, each neuron contains an operational parameter, 3 hidden layers, and an output layer with one output neuron. The model achieved an average of 98% accuracy and high precision, recall, and $F1$-Score.

Another DL technique is called long short-term memory (LSTM). Tian et al. [29] developed a LSTM for monitoring the status of reciprocating compressor place on offshore platforms. The components tested included bearings, motors, and valves. They were able to achieve higher prediction accuracy compared to the other baseline models and achieved significant reduction in different error metrics.

2 System Description and Methodology

2.1 Overview

This section explains the steps taken to build and train a fault multi-label classification model for predictive maintenance of reciprocating compressors. It starts with describing the dataset used in the study, which includes how the data is collected and labeled. Then comes the pre-processing part which shows how the data was cleaned and prepared for training. After that, splitting and normalization process is described. This is followed by an explanation on the structure of the model, training process, and how the model is evaluated using different metrics. It ends by explaining the Bayesian optimization technique and how the model will be tuned. This section covers step-by-step methodology, presented in Fig. 3, to build a prediction MLP model (Fig. 5).

2.2 Data Collection

The dataset used in this methodology comes from an open-source dataset made for predictive maintenance for reciprocating compressors. The data provider [18] collected several thousands of samples from an air compressor system and filtered out the most useful 1000 rows for analysis. These rows cover different parameter conditions that will help with fault classification.

The dataset contains sensor measurements collected from a single stage, two-cylinder, water-cooled reciprocating air compressor. The compressor supplies air to a factory line and is driven by an AC electric motor which can produce up to 8 bars of compressed air (Table 3).

The dataset also contains 5 columns of target labels that tell the health status of 5 key components in the compressor which are the discharge valve, radiator, bearings, water pump, and AC motor. The labels include terms like "stable," "ok," "noisy," "clean," and "dirty." These terms will be eventually replaced into binary formats such as 0 for healthy and 1 for faulty for the purpose of fault classification. Since the dataset is filtered from a larger sample of data, each row is treated as an individual snapshot taken during a specific test condition.

2.3 Data Pre-processing

The first step was to import the dataset and check for any duplicate rows and missing values. No duplicate rows were found which means each row has a different operating condition. There also was no null data present so no further action to fill in or remove rows were needed.

Fig. 3 Workflow diagram of the methodology

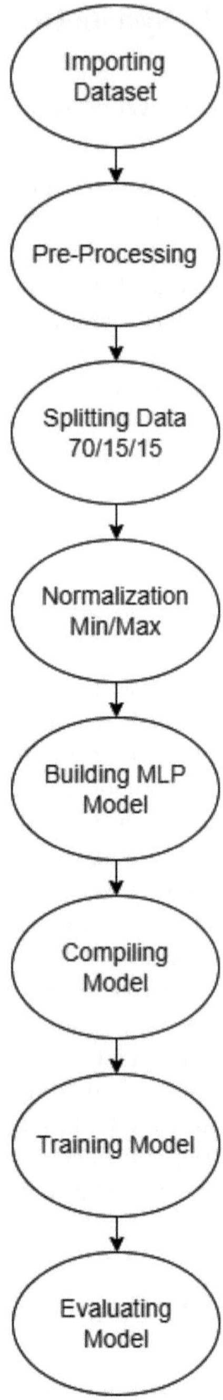

Table 3 Input parameters

Input features	Description
RPM	Number of rotations per minute of the motor
Motor power	Power consumption of the motor in (kW)
Torque	Torque provided by motor (Nm.)
Discharge pressure	Pressure leaving the compressor in (bar)
Air flow	Flow rate of air into the compressor (cubic meters per minute)
Noise	Noise level of the compressor (dB)
Discharge temperature	Temperature of the air leaving (°C)
Water pump outlet pressure	Outlet pressure of the water pump (bar)
Water inlet temperature	Inlet temperature of the cooling water (°C)
Water outlet temperature	Outlet temperature of the cooling water (°C)
Water pump power	Power consumption of the water pump (kW)
Water flow	Flow of the cooling water (cubic meters per minute)
Oil pump power	Power consumption of the oil pump (kW)
Oil tank temperature	Temperature of the oil in tank (°C)
Ground acceleration	Acceleration on the foundation of the compressor (X, Y, Z direction)
Head acceleration	Acceleration on the head of the compressor (X, Y, Z direction)

It was also noticed that the health state of the ac motor was stable across all the rows in the dataset. Since this health condition has no effect on the fault detection, it was dropped from the dataset.

Each component in the dataset had a different set of labels to describe their health status. The radiator and discharge valve were labeled as either "clean" or "dirty," while the bearings and water pump were labeled either "ok" or "noisy". These output labels were analyzed to check how they were distributed across the four components to identify how many samples were marked as healthy and faulty to check for imbalances that might affect the training process.

As shown in Fig. 4, each of the four components contain 800 healthy samples and 200 faulty samples, meaning that there is an imbalance in the dataset which should be taken into consideration.

The dataset was then divided into numerical features and non-numerical labels. The numerical features are kept aside to be normalized after splitting the data into training, validation, and testing sets. Each non-numerical label was then converted into binary values to make more suitable for fault classification. The terms "ok" and "clean" were both given the value of 0 to represent a healthy condition, while "noisy" and "dirty" were given the value 1 to represent a faulty condition.

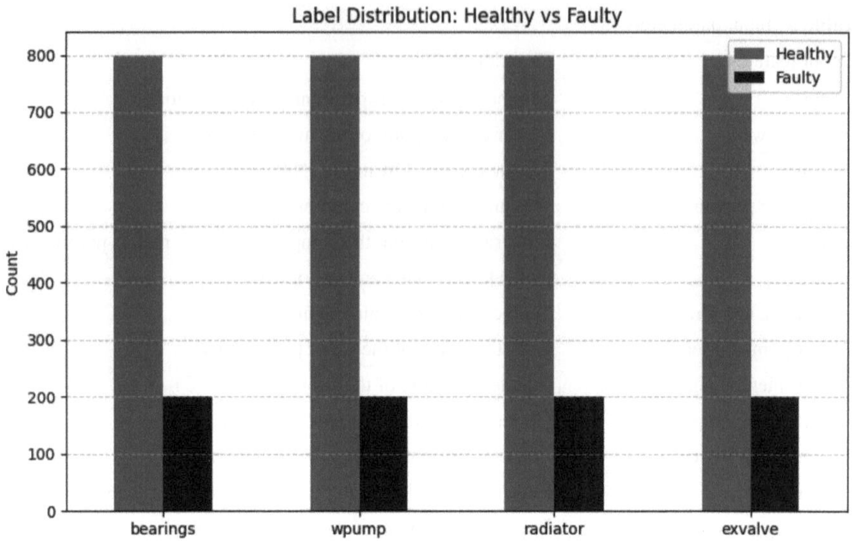

Fig. 4 Distribution of healthy and faulty states per component

2.4 *Splitting and Normalization*

After pre-processing, the data was split into training, validation, and testing sets to train the model and evaluate the performance. A 70-15-15 split was used, where 70% of that is used for training, 15% for validation, and 15% for testing. This approach allows the model to learn from most of the data, prevent overfitting, and see how the model performs on unseen data.

The numerical data was then normalized using Min–Max, so the values are scaled between 0 and 1. This is done to make the learning process more efficient. The scaling was fitted into the training dataset and then applied to both the validation and testing sets.

$$X_n = \frac{X - X_{\min}}{X_{\max} - X_{\min}} \tag{1}$$

where X_n is the normalized value, X is the original value, and $X_{\max}, X_{\min}$ are the maximum and minimum values in that column.

2.5 *Model Architecture*

In this study, an MLP was designed to classify the health state of four components in a reciprocating compressor. An MLP typically consists of an input layer, one or more hidden layers, and an output layer. Each layer is fully connected to the next layer,

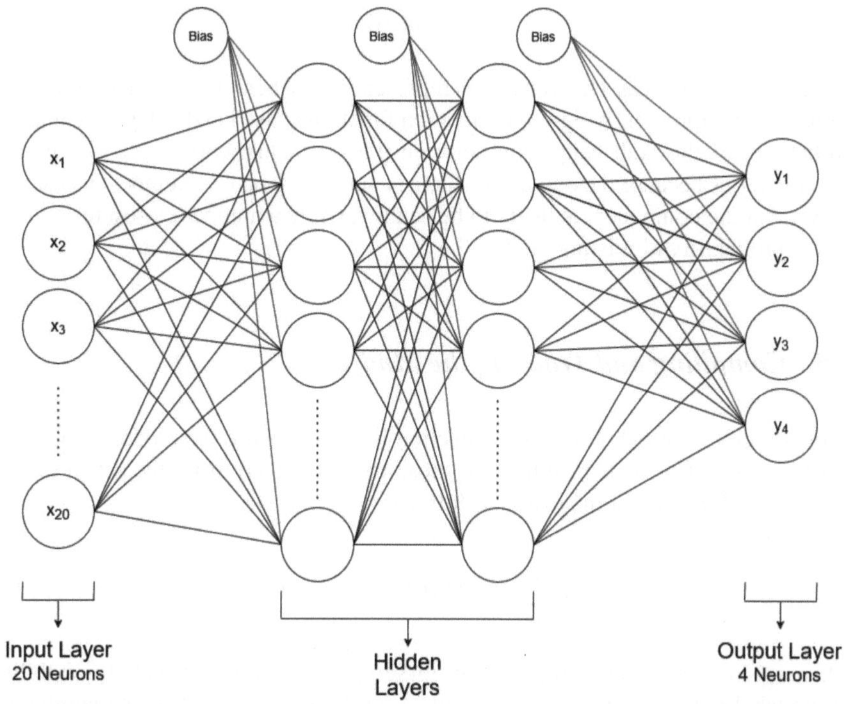

Fig. 5 Model architecture

meaning that each neuron in one layer feeds its output to every neuron in the next layer [24]. The architecture of the methodology model consists of an input layer, two hidden layers, and an output layer.

The input layer is made up of 20 neurons that each represents a normalized sensor measurement from the compressor system. Each input is passed to the first hidden layer (Fig. 5).

The first hidden layer contains 41 neurons, and the second layer contains 21 neurons. The number of neurons were chosen based on Eq. 2, where n is the number of neurons in the input layer, for the purpose of defining a starting point [16]. In a hidden layer, each neuron performs a weighted sum of its inputs, adds a bias, and passes it to an activation function.

$$h = 2s + 1 \tag{2}$$

Both hidden layers use the Rectified Linear Unit (ReLU) activation function which helps the model learn complex patterns and correlations. ReLU is used because it has been shown to speed up the convergence in deep learning models and solves the issue of vanishing gradients that can be seen in other activation functions such as sigmoid and tan$h$ [10].

$$f(x) = \max(0, x) \tag{3}$$

The output layer has 4 neurons, each representing the health status of a component within the compressor. A sigmoid activation function is used in this layer to map the probability of the fault happening between 0 and 1. This works well with binary classifications because the outputs are already set to be 0 or 1. A number closest to 0 shows that the component is likely to be healthy, while a number closest to 1 shows a high probability of it being faulty.

2.6 Compiling and Training the Model

A binary cross-entropy (BCE) loss function is used because the model outputs are binary values. They calculate how far those predictions are from the actual binary values. The formula for the loss is as follows:

$$\text{BCE} = -\frac{1}{N}\sum_{i=1}^{N}[y_i \log(p_i) + (1 - y_i)\log(1 - p_i)] \tag{4}$$

where y_i is the actual label (0 or 1) for the ith sample and p_i is the predicted probability of the label being faulty (between 0 and 1) for the ith sample. N is the number of samples.

After the loss is calculated, a backward pass is performed where each gradient is used to adjust the weights and biases with the help of Adam optimizer. This is repeated for every epoch until the model achieves an acceptable performance. After training, loss, accuracy, recall, and precision were plotted to visualize the performance of the model per epoch.

2.7 Model Evaluation

The model is then used to make predictions on the testing dataset. The model outputs probabilities between 0 and 1 for each of the four component conditions. These probabilities are converted into binary values using a threshold of 0.5. Any value above 0.5 is considered faulty, while anything below is considered healthy. To see how well the model performed, the accuracy per label and row were checked. The predictions are also evaluated using precision, recall, and $F1$-Score metrics (Table 4).

To improve the performance of the model, Bayesian optimization was used to fine tune the hyperparameters to minimize the number of trainable parameters. The defined search space included the number of neurons in the first hidden layer ranging from 16 to 32 and the second hidden layer ranging from 8 to 16. The objective function was set to be validation recall, so that the model prioritizes learning to correctly

Table 4 Metrics and their definitions

Metric	Formula	Description
Precision	$\frac{TP}{TP+FP}$ (5)	Out of all the times the model predicted faulty, how many times they were actually faulty
Recall	$\frac{TP}{TP+FN}$ (6)	Out of all the faulty components, how many did the model find
$F1$-score	$\frac{2*Precison*Recall}{Precision+Recall}$ (7)	Harmonic mean of precision and recall

classify faulty conditions since they are the minority class that account only 20% of dataset. This helps reduce the failures that are predicted to be healthy, which is important in predictive maintenance where missed faults can lead to equipment failure. At the end, the top 10 models ranked by their validation performance and the model that offered a balance between performance and complexity was chosen for final evaluation.

3 Results and Discussion

3.1 Training Results

The model was trained in over 100 epochs, and the performance was tracked across both training and validation datasets using the following metrics:

3.1.1 Accuracy Per Epoch

Figure 6 shows the training (blue line) and validation (red line) accuracy in over 100 epochs. In the initial phase, both training and validation accuracy start low and followed by a rapid increase up to 0.96–0.97 until it starts to stabilize around 35th epoch. It can be noticed that the validation accuracy plateaus are a bit lower around 0.92. This means that the model generalizes well on the validation set without overfitting.

3.1.2 Loss Per Epoch

Figure 7 shows the training (blue line) and validation (red line) loss over 100 epochs. At the initial stages, both lines start high around 0.65 and decrease almost similarly as training continues and they start converging near zero showing that the model is learning without overfitting.

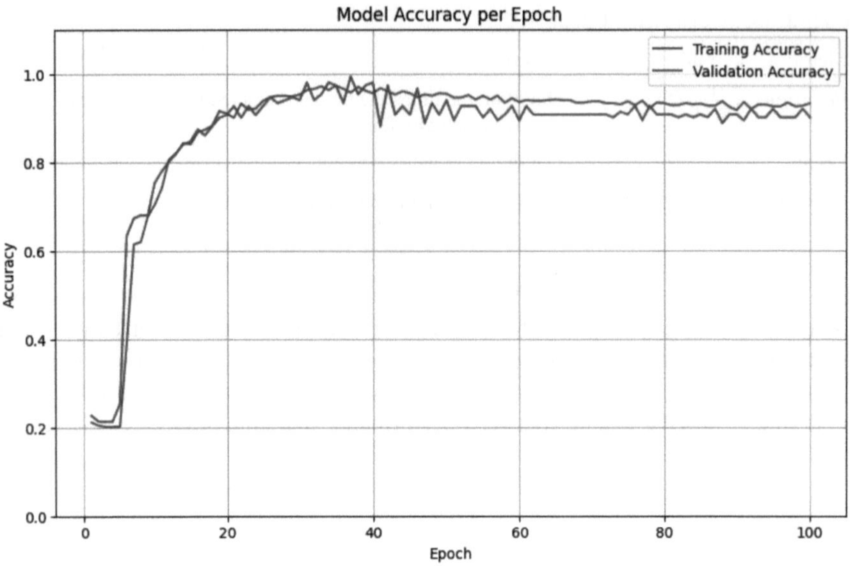

Fig. 6 Accuracy per epoch

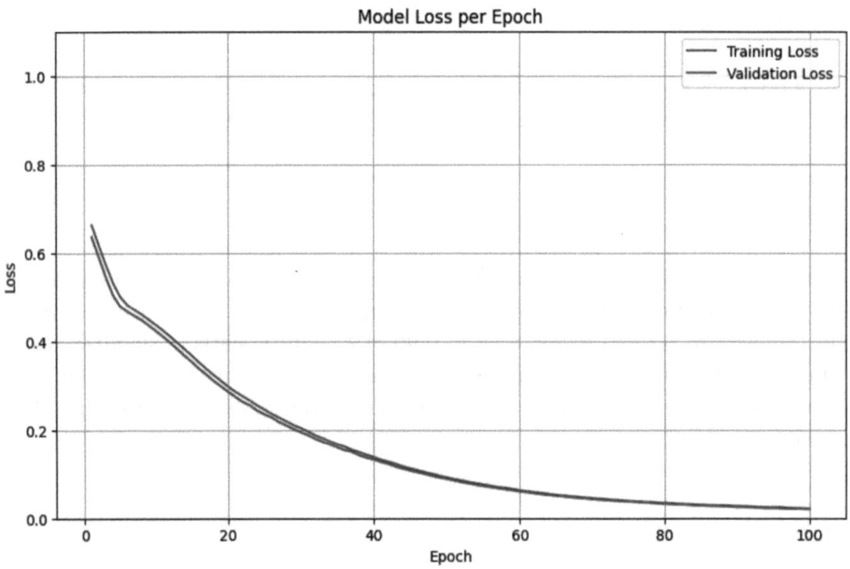

Fig. 7 Loss per epoch

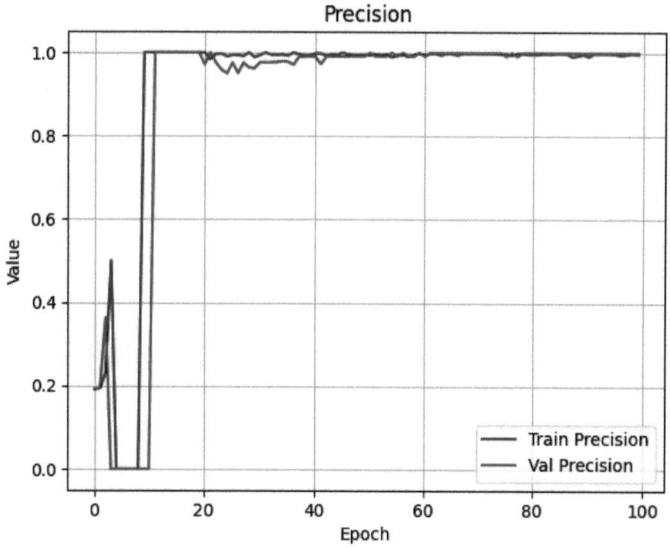

Fig. 8 Precision per epoch

3.1.3 Precision Per Epoch

Figure 8 shows training (blue line) and validation (red line) precision changed over 100 epochs. As can be seen initially, there are fluctuations that are followed by a steep drop to zero. This is because the model is still learning and has made a few wrong predictions. As the training progressed, the precision quickly improved and stabilized around 1.0. This is a sign that the model got much better at predicting faults without generating many fault positives. The validation precision closely followed a similar pattern, showing the model performs well on unseen data.

3.1.4 Recall Per Epoch

Figure 9 shows how training (red line) and validation (blue line) recall changed over 100 epochs. Like precision, recall was low at the beginning because the model is not able to detect many of the faulty cases yet. After about 15 epochs, recall started improving quickly as the model got better at catching the faulty samples and as it reached the 50th epoch, both training and validation approached close to 1.0 and stayed steady up until the last epoch.

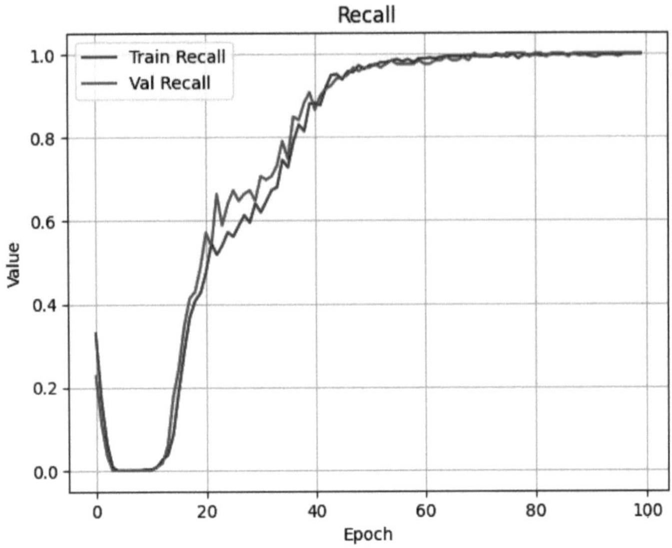

Fig. 9 Recall per epoch

3.2 *Evaluating the Testing Dataset*

After training, the final model was evaluated on the testing dataset to determine the performance of the model on unseen data. The predictions by the model were compared with the actual labels, and the accuracy per label and rows were measured.

$$\text{Accuracy}_{\text{per label}} = \frac{\text{Number of Correct Predicitions for that label}}{\text{Total Number of Predictions for that label}} \tag{8}$$

$$\text{Accuracy}_{\text{per row}} = \frac{\text{No. of Samples where all labels are predicted}}{\text{Total number of samples}} \tag{9}$$

The model achieved perfect accuracy on the water pump, radiator, and discharge valve, meaning that it predicted their health status correctly for every sample in the testing dataset. Bearing also showed 98% accuracy, missing only a few classifications (Fig. 10).

Measuring accuracy alone can be misleading when there is an imbalance in the data. Since most of the data are healthy samples (80%) and only a few are faulty (20%), the model could get a high accuracy just by ignoring the faults and guessing healthy states for everything. That is why other metrics like precision, recall, and $F1$-Score were also used to evaluate the performance of the model (Tables 5 and 6).

The model achieved perfect precision and recall on the water pump, radiator, and discharge valves. The model still performed well in the case of bearing even though the model had a lower precision (0.93) and recall (0.96) compared to the

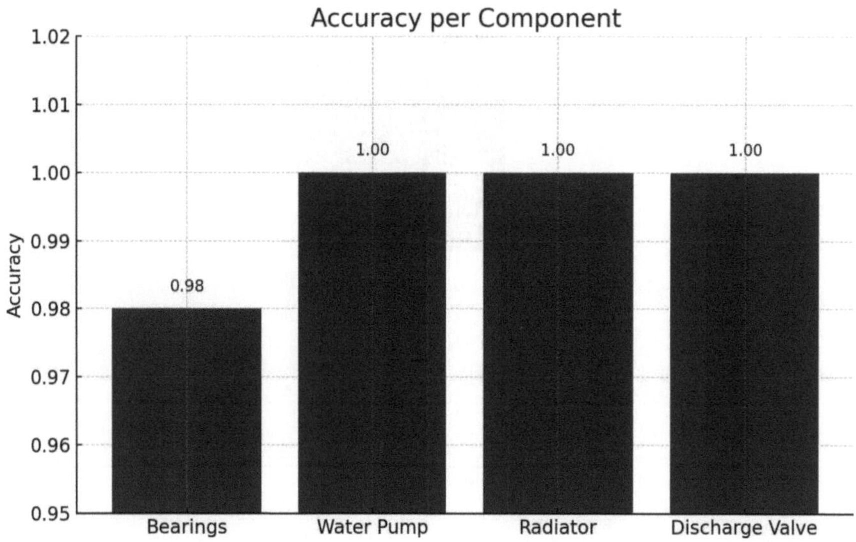

Fig. 10 Accuracy per label

Table 5 Accuracy metrics

Metric	Value
Average accuracy per label	0.995
Accuracy per row	0.98

Table 6 Classification report

Components	Precision	Recall	$F1$-score	Support
Bearings	0.93	0.96	0.95	28
Water pump	1.0	1.0	1.0	26
Radiator	1.0	1.0	1.0	27
Discharge valve	1.0	1.0	1.0	32

other components. Overall, the model did pretty well, and the results show that it can classify faulty conditions even though they only account for 20% of the total data.

Figure 11 shows the confusion matrix for bearings that explains how the model made the right and wrong classifications. The model predicted 120 healthy bearings and 27 faulty statuses correctly. Only 2 times the model classified the condition as faulty when it was actually healthy, and one time it missed a faulty condition classifying it as healthy.

This shows that the model performed very well at predicting both conditions with some minor mistakes. But overall, it is reliable because most of the predictions were correct, and sometimes in predictive maintenance, it is better to have false alarms

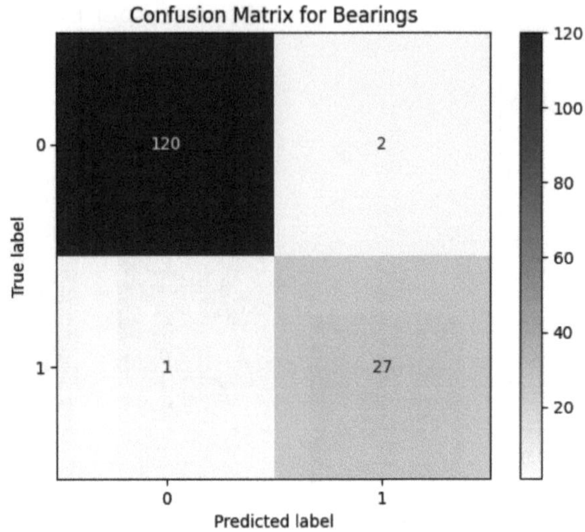

Fig. 11 Bearings confusion matrix

than to actually miss a real fault but still missing one fault means that there is room for improvement.

3.3 Optimization and Model Tuning

To tune the model, Bayesian optimization was used to search for the best hyperparameters. Before tuning the model, the trainable parameters were 1831. The main goal for optimizing the model was to find the best combinations of the number of neurons in the hidden layers to potentially lower the number of trainable parameters. Initially, the objective for the search was set to be validation accuracy. This showed how misleading the accuracy metric can be for multi-classification models that include data imbalances. Telling the model to prioritize validation accuracy meant that the model only cared about getting a high accuracy without focusing on the minority that is the faulty classifications. The model would have high accuracy but as the precision and recall for each label is measured, both those metrics would come back as zeros, meaning that the model was ignoring faulty conditions by just predicting a healthy state for every classification. When validation recall was used instead of the accuracy metric, the model would prioritize the faulty samples more during the training. Even though the overall accuracy of the model was lower, the model was identifying the faults properly.

The top 10 models were chosen from the search and were evaluated based on the balance between the performance measured using the macro $f1$ metrics and the number of parameters (Table 7).

Table 7 Top 10 best models

Model index	Params	HL1	HL2	Macro $F1$-score
41	972	24	16	0.991379
17	1194	30	16	0.991379
97	972	24	16	0.991071
7	1194	32	14	0.991071
10	1268	32	16	0.991071
77	972	24	16	0.987103
16	1194	30	16	0.987103
38	898	22	16	0.986842
36	914	30	8	0.986842
1	1124	30	14	0.986842

As shown in Table 8 and Fig. 12, the best model had 972 parameters, which is almost half compared to the parameters in the initial model that had 1831 parameters. The tuned model is much smaller and more efficient to train.

The number of neurons in each hidden layer is (24, 16), when initially it was (41, 21). The performance measured using the macro $F1$-Score metric has improved to 0.99137 from it initially being 0.98275. This shows that Bayesian optimization not only helped find a better architecture for the model but also improved the performance.

Table 8 Comparison of initial and tuned model

Model	Trainable parameters	Hidden layer 1	Hidden layer 2	Macro $F1$-Score
Initial model (before tuning)	1831	41	21	0.98275
Best model	972	24	16	0.99137

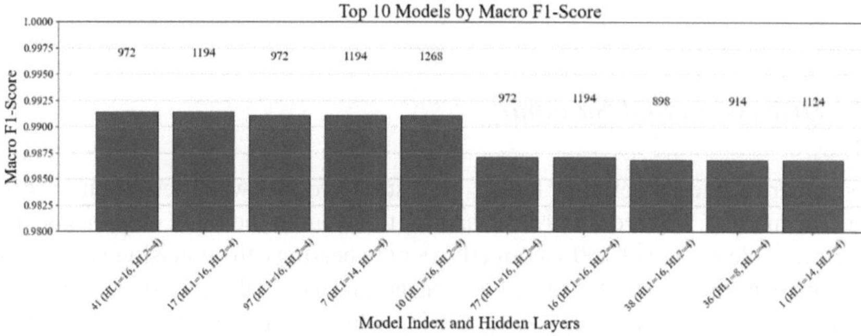

Fig. 12 Comparison of the top 10 models

Table 9 Best model performance on testing dataset

Components	Precision	Recall	$F1$-Score	Support
Bearings	0.93	1.0	0.97	28
Water pump	1.0	1.0	1.0	26
Radiator	1.0	1.0	1.0	27
Discharge valve	1.0	1.0	1.0	32

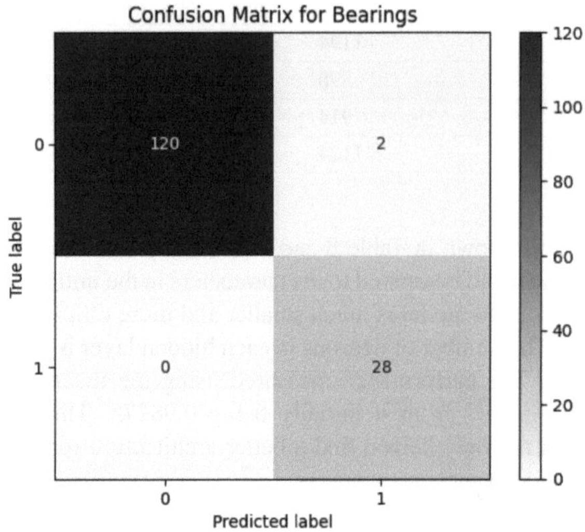

Fig. 13 Bearing confusion matrix for the best model

Table 9 shows the results of the best model when tested on the testing dataset. For all the components, except for bearing, the model was able to predict their health status 100% of the time, not missing any faults and wrong classifications. For the bearing component, the model has a lower precision of 0.93 but the recall is 1.0. This means that the model was able to correctly identify all the faulty states but also classified two healthy states as faulty states. This can be seen in Fig. 13, which shows a visual representation of all the predictions against the true labels.

3.4 Discussion and Summary

To summarize, the model after tuning gave strong results when tested on unseen data. The macro $F1$-Score is 0.99137, and for three out of the four components, the model was able to predict 100% correctly. For the bearings, the precision is high but is lower when compared to the other components and the recall was 1.0, showing that the model never missed a faulty state. Having a high recall for all the components is most important in predictive maintenance as the failures should not go unnoticed, causing catastrophic failures, unplanned downtime, and high maintenance costs. The Bayesian optimization tuning process helped reduce the number of parameters by almost half and improved the fault classification. Even though the model performed well, there were still false classifications as some healthy states were predicted to be faulty, this means that the model still has room for improvement. Also, the possibility of using one hidden layer can give better results as the number of parameters will be reduced more. Trying a simpler architecture with one hidden layer can be a good idea to explore in the future.

4 Conclusion and Future Work

In this thesis, a data-driven approach was used to build a fault multi-label classification model for predictive maintenance of reciprocating compressors. The goal was to predict the health conditions of four key components: radiator, bearing, water pump, and discharge valve using snapshot data. An MLP model was built, trained, tuned, and evaluated on air compressor data. The tuned model achieved 99% average macro $F1$-Score and 100% recall on all components. This is important in predictive maintenance as undetected faults that are classified as healthy can cause serious damage and high costs.

Bayesian optimization was used to tune the hyperparameters, which helped reduce the number of trainable parameters by almost half while also improving the performance of the model. Although the results were strong, there were some false alarms where healthy components were labeled faulty by the model which shows that there is still room for improvements. Perhaps trying a simpler architecture with one hidden layer in the future can improve the performance of the model.

The dataset used in this study is limited and filtered, with the labels being general rather than including more specific fault types. This does not reflect the real-world conditions and limits how detailed the predictions of the models can be.

The study also explored the importance of condition monitoring, with a focus on snapshot monitoring, to assess the health state of reciprocating compressors. Traditional methods of monitoring depend on detecting the anomalies in the compressor and then manually comparing them with the reference data to classify faults. Even though this model is a proof of concept, it still shows the potential of how machine learning can be used one day to help operators with automated predictions and help with their decision-making.

For future work, this approach should be tested on real compressor systems to understand how well it can perform in a practical setting, especially focusing on valves, since they are the most common cause of failure shutdowns, using more detailed labels such as leakage, spring failure, or mechanical wear to build a more realistic model.

Overall, this thesis shows that data-driven fault classification is possible on snapshot data. It has the potential to shift from traditional monitoring techniques to better predictive strategies.

Appendix 1: Data Pre-processing

Find pre-processing code in the next page.

```
# Importing Libraries
import pandas as pd
import matplotlib.pyplot as plt
import numpy as np
import tensorflow as tf
from tensorflow import keras
from tensorflow.keras.metrics import Precision, Recall, AUC
from sklearn.preprocessing import MinMaxScaler
from sklearn.model_selection import train_test_split
from sklearn.metrics import classification_report, confusion_matrix,
f1_score, ConfusionMatrixDisplay


# Improting Dataset
file_path = '/content/drive/MyDrive/Thesis/data (1).csv'


# Separating Numerical Data
numeric_columns = df.select_dtypes(include=['number']).columns
df_numeric = df[numeric_columns]

# Sepearting Non-numerical Data
non_numeric_columns = df.select_dtypes(exclude=['number']).columns
df_non_numeric = df[non_numeric_columns]


# Converting Classification Labels into Binary Values (0, 1)
df_non_numeric['bearings'] = df['bearings'].map({'Ok': 0, 'Noisy': 1})
df_non_numeric['wpump'] = df['wpump'].map({'Ok': 0, 'Noisy': 1})
df_non_numeric['radiator'] = df['radiator'].map({'Clean': 0, 'Dirty': 1})
df_non_numeric['exvalve'] = df['exvalve'].map({'Clean': 0, 'Dirty': 1})
df_non_numeric


# Splitting Features (Inputs) and Labels (Output)
input_data = df_numeric
output_data = df_non_numeric

# Splitting Data into Training (70%) and Temporary (30%)
input_train, input_temp, output_train, output_temp =
train_test_split(input_data, output_data, test_size = 0.3, random_state =
1697639)

# Spliting Temporary Data (30%) into Validation (15%) amd Testing (15%)
input_val, input_test, output_val, output_test =
train_test_split(input_temp, output_temp, test_size = 0.5, random_state =
1697639)
```

Appendix 2: Model Architecture

```
# Architecture of the Neural Network Model
model = keras.Sequential([
    keras.layers.Input(shape = (20,)),
    keras.layers.Dense(41, activation = 'relu'),
    keras.layers.Dense(21, activation = 'relu'),
    keras.layers.Dense(4, activation = 'sigmoid')
])

# Compiling the Model
model.compile(optimizer ='adam', loss ='binary_crossentropy', metrics
=  ['accuracy', Precision(name = 'precision'), Recall(name = 'recall'),
AUC(name = 'auc')])

# Saving training history
history = model.fit(training_ds, epochs = 100, validation_data =
validation_ds)
```

Appendix 3: Hyperparameter Tuner

```
def build_model(hp):
  model = keras.Sequential([
    keras.layers.Input(shape = (20,)),
    keras.layers.Dense(units = hp.Int('units_layer1', min_value = 16,
max_value = 32, step = 2), activation = 'relu'),
    keras.layers.Dense(units = hp.Int('units_layer2', min_value = 8,
max_value = 16, step = 2), activation = 'relu'),
    keras.layers.Dense(4, activation = 'sigmoid')
  ])

# Compiling the Model
 model.compile(optimizer ='adam', loss = 'binary_crossentropy', metrics =
[Precision(name = 'precision'), Recall(name = 'recall'), AUC(name =
'auc')])

  return model


tuner = kt.BayesianOptimization(
    build_model,
    objective='val_recall',
    max_trials= 150,
    executions_per_trial= 2,
    directory='MODELS40',
    project_name='bayesian_tuning'
)

# Run the search
tuner.search(training_ds, epochs=100, validation_data= validation_ds,
verbose = 0)
```

References

1. Ahmed M, Gu F, Ball AD (2012) Fault detection of reciprocating compressors using a model from principles component analysis of vibrations. J Phys: Conf Ser 364(1). https://doi.org/10.1088/1742-6596/364/1/012133
2. Al-Badour F, Sunar M, Cheded L (2011) Vibration analysis of rotating machinery using time–frequency analysis and wavelet techniques. Mech Syst Signal Process 25(6):2083–2101. https://doi.org/10.1016/j.ymssp.2011.01.017
3. Bisal P, Jana P (2025) Compressor health monitoring and predictive maintenance by anomaly detection using fuzzy logic and random forest model by machine learning process through Matlab. Int J Mach Tools Maint Eng 6(1). https://www.researchgate.net/publication/389045908
4. Bloch H (2006) A practical guide to compressor technology
5. Bloch H, Hoefner J (1996) Reciprocating compressors operation and maintenance
6. Corvaro F, Giacchetta G, Marchetti B, Recanati M (2017) Reliability, availability, maintainability (RAM) study, on reciprocating compressors API 618. Petroleum 3(2):266–272. https://doi.org/10.1016/j.petlm.2016.09.002
7. Gong W, Chen H, Zhang Z, Zhang M, Wang R, Guan C, Wang Q (2019) A novel deep learning method for intelligent fault diagnosis of rotating machinery based on improved CNN-SVM and multichannel data fusion. Sensors 19(7):1693. https://doi.org/10.3390/s19071693

8. Griffith WA, Flanagan EB (2001) Online, continuous monitoring of mechanical condition and performance for critical reciprocating compressors. Texas A&M University. Turbomachinery Laboratories, pp 209–218. https://oaktrust.library.tamu.edu/items/29caf0ea-873e-4f09-ad3c-78df6e21cd4d
9. Guerra C (2013) Condition monitoring of reciprocating compressor valves using analytical and data-driven methodologies. http://scholarworks.rit.edu/theses
10. Guo FY, Zhang YC, Wang Y, Ren PJ, Wang P (2021) Fault diagnosis of reciprocating compressor valve based on transfer learning convolutional neural network. Math Prob Eng 2021https://doi.org/10.1155/2021/8891424
11. Kolodziej JR, Trout JN (2018) An image-based pattern recognition approach to condition monitoring of reciprocating compressor valves. J Vib Control 24(19):4433–4448. https://doi.org/10.1177/1077546317726453
12. Leonard SM (1997) Increase reliability of reciprocating hydrogen compressors. Hydrocarb Process 75(1)
13. Leoni L, De Carlo F, Abaei MM, BahooToroody A, Tucci M (2023) Failure diagnosis of a compressor subjected to surge events: a data-driven framework. Reliab Eng Syst Saf 233https://doi.org/10.1016/j.ress.2023.109107
14. Lv Q, Yu X, Ma H, Ye J, Wu W, Wang X (2021) Applications of machine learning to reciprocating compressor fault diagnosis: a review. In: Processes, vol 9, no 6. MDPI AG. https://doi.org/10.3390/pr9060909
15. Mobley RK (2002) An introduction to predictive maintenance—R. Keith Mobley—Google Books. https://books.google.com.om/books?hl=en&lr=&id=SjqXzxpAzSQC&oi=fnd&pg=PP1&dq=predictive+maintenance&ots=iHrONADjen&sig=FsyNuGTlHV00ypxX-ick5KGy9hk&redir_esc=y#v=onepage&q=predictive%20maintenance&f=true
16. Mohammadzaheri M, Chen L, Grainger S (2012) A critical review of the most popular types of neuro control. Asian J Control 14(1):1–11. https://doi.org/10.1002/asjc.449
17. Nwakpang IA, Lebele-Alawa BT, Nkoi B (2019) Performance assessment of a two-stage reciprocating air compressor. Eur J Eng Res Sci 4(4):74–82. https://doi.org/10.24018/ejers.2019.4.4.1130
18. Okudan A (2023) Predictive maintenance dataset—air compressor. https://www.kaggle.com/datasets/afumetto/predictive-maintenance-dataset-air-compressor
19. Prometheus Group (2021) The importance of condition based monitoring | Prometheus group. https://www.prometheusgroup.com/resources/posts/the-importance-of-condition-based-monitoring
20. Qi G, Zhu Z, Erqinhu K, Chen Y, Chai Y, Sun J (2018) Fault-diagnosis for reciprocating compressors using big data and machine learning. Simul Model Pract Theory 80:104–127. https://doi.org/10.1016/j.simpat.2017.10.005
21. Qureshi F (2022) Effectiveness of crosshead vibration measurement in capturing issues on cylinder of reciprocating compressor. Baker Hughes ORBIT ARTICLE. https://www.bakerhughes.com/bently-nevada/orbit-home/orbit-article/reciprocating-compressor-condition-monitoring-q2
22. Royce NB (2011). Reciprocating compressors introduction. https://books.google.com.om/books/about/Compressors.html?id=irfr-Kd1nvEC&redir_esc=y
23. Sánchez R-V, Macancela J-C, Cabrera D, Cerrada M (2024) Fault classification in reciprocating compressors: a comparison of machine learning and deep learning approaches. IFAC-PapersOnLine 58(8):157–161. https://doi.org/10.1016/j.ifacol.2024.08.066
24. Saturday EG, Callistus AM (2025) Development and evaluation of artificial neural network techniques for predictive maintenance of a reciprocating compressor. Int J Adv Eng Manag 7(3):371–383. https://doi.org/10.35629/5252-0703371383
25. Schirmer A (2004) Online monitoring of Recip compressors, 2004 | PDF | Gas Compressor | Monitoring (Medicine). https://www.scribd.com/document/207282116/Online-Monitoring-of-Recip-Compressors-2004
26. Schultheis S, Lickteig C, Parchewsky R (2011) Reciprocating compressor condition monitoring. In: Proceedings of the first Middle East turbomachinery symposium. https://oaktrust.library.tamu.edu/items/cb46a761-7b00-473c-9fee-4d3e6814d41b

27. Smith TJ, Schultheis SM, Hanifan M (1997) Morris plant improves compressor reliability and Onstream time using reciprocating compressor monitoring
28. Stewart M (2019) Reciprocating compressors. In: Surface production operations. Elsevier, pp 655–778. https://doi.org/10.1016/B978-0-12-809895-0.00009-0
29. Tian H, Li R, Yang L (2022) Operation status monitoring of reciprocating compressors based on the fusion of spatio-temporal multiple information. Meas: J Int Meas Confed 204. https://doi.org/10.1016/j.measurement.2022.112087
30. Zhao H-Y, Wang J-D, Xing J-J, Gao Y-Q (2015) 774. A feature extraction method based on LMD and MSE and its application for fault diagnosis of reciprocating compressor. 17(7)
31. Zhu T, Ran Y, Zhou X, Wen Y (2019) A survey of predictive maintenance: systems, purposes and approaches. http://arxiv.org/abs/1912.07383

Development of AI Model to Estimate the Head of an Electrical Submersible Pump for Liquids with Various Viscosities

Asifur Rahman, Morteza Mohammadzaheri, and Ali Al-Humairi

Abstract The performance curves of electrical submersible pumps (ESPs) are commonly characterised using water as the working fluid by their manufacturers. However, ESPs are widely deployed for oil production where the fluid viscosity is significantly higher than water, and conventional methods of head estimation can become unreliable. Analytical methods rely on complex derivations, empirical methods may have limited applicability, and computational fluid dynamics (CFD) requires heavy computational resources. This paper explores the applications of artificial intelligence (AI), specifically artificial neural networks (ANNs) to develop fast and accurate models for predicting head of an ESP under varying viscosities. Two ANN architectures,a multilayer perceptron (MLP) and a fully connected cascade (FCC) network, were developed and trained on experimental data provided by (Shi et al. in Proc Inst Mech Eng, Part A: J Power Energy 235:1976–1991, 2021 [25]). The experimental data consisted combinations of various motor speeds, viscosities and flowrates with their corresponding head. The methodology included data preprocessing, development of preliminary models and hyperparameter optimisation using random search and Bayesian optimisation. Both models demonstrated strong prediction metrics, with a mean absolute error (MAE) below 0.6 ft and mean absolute percentage errors under 8% on the unseen testing dataset. Whilst the FCC model outperformed the MLP model in terms of raw metrics, the MLP model demonstrated smoother and more traditionally expected head curves. The MLP model also requires fewer number of parameters making them easier to train and more suitable for deployments. The study however uncovered an unexplained sensitivity of the MLP model to the normalisation technique used for the output data—suggesting an area for future research. Overall, this study highlights the potential of AI-based models in head predictions for ESPs.

Keywords Electrical submersible pumps · Viscous liquids · Head prediction · Artificial neural networks · Artificial intelligence

---

A. Rahman · M. Mohammadzaheri (✉) · A. Al-Humairi
German University of Technology in Oman, Halban, Sultanate of Oman
e-mail: morteza.zaheri@gutech.edu.om

A. Azizi (ed.), *Smart Diagnostics and Predictive Maintenance*, Emerging Trends in Mechatronics, https://doi.org/10.1007/978-981-95-3921-5_3

Abbreviations

ESP	Electrical Submersible Pump
GLR	Gas–Liquid Ratio
CFD	Computational Fluid Dynamics
3D	3-Dimensional
AI	Artificial Intelligence
ANN	Artificial Neural Network
MAE	Mean Absolute Error
MLP	Multilayer Perceptron
FCC	Fully Connected Cascade
BEP	Best Efficiency Point
RANS	Reynolds-Averaged Navier–Stokes
SST	Shear-Stress Transport
NNFIR	Neural Network Finite Impulse Response
SMOTE	Synthetic Minority Oversampling Technique
BPANN	Backpropagated Artificial Neural Network
CSV	Comma-Separated Value

Symbols

H_{pump}	Head developed by pump
ΔP_{pump}	Pressure developed by pump
ρ	Density of working fluid
g	Gravitational acceleration
HP_h	Hydraulic horsepower of pump
HP_b	Brake horsepower of pump
η_1	Pump efficiency from performance curve
η_2	Desired pump efficiency
Q_1	Pump flowrate from performance curve
Q_2	Desired pump flowrate
H_1	Pump head from performance curve
H_2	Desired pump head
N_1	Motor speed from performance curve
N_2	Desired motor speed
H_{Euler}	Euler pump head
H_{loss}	Head loss due to non-ideal fluid
Q_{vis}	Viscous flowrate of pump
Q_w	Water flowrate of pump
C_Q	Flowrate correction factor
H_{vis}	Viscous head of pump
H_w	Water head of pump

C_H	Head correction factor
η_{vis}	Viscous efficiency of pump
η_w	Water efficiency of pump
C_η	Efficiency correction factor
x	Point from dataset
$\overline{x}$	Mean of dataset
σ_x	Standard deviation of dataset
x'_m	Min–max scaled value
x'_z	Z-score standardised value
x'_r	Robust-scaled value
s	Sum of the weighted inputs and the bias
W_{in}	Input units in weight tensor
W_{out}	Output units in weight tensor
n	Number of data points
Y_i	Experimental data pint
$\hat{Y}_i$	Predicted data point
R^2	Coefficient of determination
SS_{res}	Residual sum of squares
SS_{tot}	Total sum of squares
e	Euler's number
h	Point in hyperparameter space
h^*	Best point in hyperparameter space
f	Objective function
p	Surrogate function
$\mathcal{D}$	Observations
$\mathcal{GP}$	Gaussian process
m	Mean function
K	Covariance
a_{UCB}	Upper confidence bound acquisition function
β	Exploration–exploitation trade-off
μ	Surrogate model's mean function
σ	Surrogate model's standard deviation function

1 Introduction

1.1 Electrical Submersible Pumps

Electrical submersible pumps (ESPs) are a form of artificial lift method that is widely used in the oil and gas industry. They are multistage centrifugal pumps that can be used to produce moderate to high volumes of fluids from oil wells. This is achieved by providing additional energy or lift to fluids with insufficient pressure inside the wellbores and hence increasing the rate of production [19].

As their name suggests, ESPs are designed to operate submerged in fluids and is powered by an electrical motor. ESPs were originally developed to work with water and were used to dewater mines and ships. However, in the following decades of their first usage, they found their application in production of oil, particularly with fluids with high water cuts and low GLR [9].

1.2 Viscosity Effects on ESP

In comparison with water, viscous fluids profoundly alter the behaviour of the centrifugal pump of an ESP. Numerical study performed by Sirino et al. [26] noted that affinity laws are only applicable for low viscosity liquids such as water and fail for oil. This implies that one cannot simply scale water curves to predict the performance under viscous conditions—as the best efficiency point shifts with viscosity.

One of the most comprehensive recent studies was carried out by Monte Verde et al. [18] where they experimentally tested six ESP models over 11 viscosities and multiple motor speeds, creating a database of more than 5800 operating conditions. They found a sharp decrease in efficiency as viscosity of the fluid increased—a monotonic performance degradation. An increase in viscosity led to a fall in head and rise in power consumption. This can be seen in their figures of head against flowrates where higher viscosity head curve shifts downwards and is also unable to achieve the same flowrates compared to that of water.

Monte Verde et al. [18] also showed that the efficiency and flowrate correction factors for viscous fluids are near independent of the flowrate in the recommended operating range. However, this does not apply to the head correction factors which do vary with flowrates.

Another noteworthy observation from [18] was that viscous losses can be mitigated with higher motor speeds and that pumps with higher specific speeds also suffer less from head loss of high-viscosity fluids. This confirms the observation from CFD study carried out by Sirino et al. [26], where they noted that head values degrade not just due to viscosity and high flowrates but also due to reduced rotational speeds.

Contrary to [12, 18] reported a non-monotonic head curve through CFD simulations of a centrifugal pump working with viscous oil. When the viscosity is in moderate range, the pump head suddenly increased above its water value. The author calls this the 'sudden rising head effect' and attributes it to boundary-layer transitions. The effect is only observed when the wall roughness is high enough with appropriate Reynolds number, which causes the viscous oil to shift from a hydraulically rough flow regime to a hydraulically smooth flow regime, thus boosting the head. With smooth wall conditions, this effect disappears and produces a monotonic head decay with increasing viscosity, too small or too large Reynolds number causes either fully smooth or rough flow regimes causing the sudden rise in head to disappear. ESPs in experimental tests do not seem to encounter this phenomenon under their normal operating conditions, and it is only observed in CFD simulations highly conditioned on high wall roughness and certain Reynolds numbers [18, 25].

Ofuchi et al. [21] conducted CFD study on multistage ESPs and observed that viscosity does not affect all the stages evenly. The first stage had a head curve that shows a deviation compared to the subsequent stages and is responsible for the largest head gain. Such behaviour is also observed in the study conducted by Stel et al. [27]. The early stages, especially the first two stages, of a multistage system act like mixers and dampen viscous losses by homogenising the flow for the subsequent stages, whilst incurring the most efficiency drop.

Across studies, there is a clear consensus that an increase in viscosity leads to a drop in pump efficiency [11, 18, 21, 26]. Interestingly, Torabi and Nourbakhsh [31] showed that even though hydraulic efficiency drops with viscosity, thick oil yields a better volumetric efficiency in a low specific speed pump. Thick oil tends to prevent internal leakage through wear rings and balance holes, which leads to this volumetric efficiency. Even though this improves volumetric efficiency, the reduced leak is completely overwhelmed by the mechanical losses and causes 'significant fall in the disc and overall efficiency'. Viscous oil is subjected to much greater drag on the rotating parts, specifically the disc friction, and the energy dissipation caused by this grows way larger causing a net drop in efficiency.

Li [11] extracted slip factors from CFD simulations and found it to be much less sensitive to viscosity compared to the flowrate. This means that the head losses, influenced by the slip factor, rely more on flow coefficient, with viscosity having a much smaller effect.

1.3 Conventional Approaches to Head Prediction

Pump manufacturers are responsible for providing performance curves of the ESPs they commercially sell. These performance curves are developed by testing the pumps on water as the working fluid at a constant motor speed. During the testing phase, they measure the flowrate, intake pressure, discharge pressure and brake horsepower. The pressure change between the intake and discharge is used to calculate the head given by Eq. (1). Pressure change along with the flowrate can be used to calculate the hydraulic horsepower given by Eq. (2). This hydraulic horsepower can be used to calculate the efficiency by finding the ratio of it to the brake horsepower, as shown in Eq. (3) [19, 30].

$$H_{\text{pump}} = \frac{\Delta P_{\text{pump}}}{\rho g} \tag{1}$$

$$\text{HP}_h = 1.67 \Delta P_{\text{pump}} Q \tag{2}$$

$$\eta = \frac{HP_h}{HP_b} \tag{3}$$

In Eqs. (1) and (2), ΔP_{pump} is the pressure gained by water during pump operation. In Eq. (1), this is measured in Pa, but in Eq. (2), this is measured in bar. In Eq. (1), H_{pump} is the head gained by the water measured in m, ρ is the density of water in kg/m^3, and g is the gravitational acceleration in m/s^2. In Eq. (2), Q is the flowrate measured in L/min, and HP_h is the hydraulic horsepower measured in W. In Eq. (3), η is the efficiency, and HP_h and HP_b are hydraulic horsepower and brake horsepower, respectively—measured in W.

The calculated head and efficiency, along with the measured brake horsepower and flowrate, are used to produce the performance curve of for a given ESP. A single graph is used to plot the head, brake horsepower and efficiency against the flowrate.

These sorts of performance curves are created for a fixed motor speed, usually at 60 Hz which corresponds to about 3500 rpm [30]. Performance curves for different motor speeds of the same pump can be developed using affinity laws [2]. For a given ESP performance curve, the new curve can be calculated using Eqs. (4), (5) and (6) to get the new flowrate, head and brake horsepower (to calculate efficiency), respectively. Note that these equations show only the relationships in terms of motor speed and ignores geometrical parameters as it is not required when working out the new values for the same pump [19].

$$\frac{Q_1}{Q_2} = \frac{N_1}{N_2} \tag{4}$$

$$\frac{H_1}{H_2} = \left(\frac{N_1}{N_2}\right)^2 \tag{5}$$

$$\frac{\text{HP}_{b1}}{\text{HP}_{b2}} = \left(\frac{N_1}{N_2}\right)^3 \tag{6}$$

In Eqs. (4), (5) and (6), N_1 and N_2 are the initial and new motor speeds, respectively, and they are measured in rpm or rad/s. Q_1 and Q_2 are the initial and new flowrate respectively, H_1 and H_2 are the initial and new head, respectively, and HP_{b1} and HP_{b2} are the initial and new brake horsepower restively.

The affinity laws are only applicable for water and other low viscous liquids and do not work for liquids with high viscosities as discussed in viscosity effects on ESP in Sect. 2. For viscous liquids, different analytical and empirical models have been developed which are discussed in the following subsections.

1.3.1 Analytical Methods

Analytical (mechanistic) models start from Euler pump head and subtract losses due to fluid viscosity. Euler pump head calculation is based on law of conservation of angular momentum. In ideal conditions, this calculation for the head varies linearly with the flowrate. However, in practice, the actual head produces a curve against the flowrate due to various viscosity effects discussed in Sect. 2 [35].

$$H = H_{\text{Euler}} - \sum H_{\text{loss}} \tag{7}$$

Equation (7) expresses the fundamental idea of mechanistic models. H_{Euler} is the ideal head based on Euler pump head calculation, the sum of H_{loss} includes different kinds of head losses caused due to viscosity effects, and H is the actual head produced by the pump.

Sun and Prado[29] created a new one-dimensional Euler equation model where they captured the qualitative viscous effects and solved the steady incomprehensible Euler equations along the pump flow path using a distributed frictional loss term.

In a more recent study by Zhu et al. [36], the authors were able to create a mechanistic model for particularly high-viscosity fluids. Their model calculates the ideal Euler head and subtracts hydraulic losses using closure relations. They also introduce the idea of a conceptual best match flowrate which is used to calculate recirculation losses due to mismatch of velocity triangles. Importantly, the model developed by them does not need new experimental calibration for each fluid—it needs the water-based performance and pump geometry. The model was validated against a large database consisting of over 200,000 data points with reasonable prediction error of 15% for medium to high flowrates.

Another recent study that uses a mechanistic model was carried out by Shi et al. [25]. They tested a multistage ESP to collect experimental data at various motor speeds, viscosities and flowrates with their corresponding heads and developed a non-ideal Euler head model. Their work validates the analytical approach of [36] and provides a very comprehensive dataset for single-phase viscous ESP flow.

Zhu et al. [33] developed a multiphase flow model using similar mechanistic ideas. Their work confirms such models can incorporate other additional physics such as gas entrainment and not just viscosity—demonstrating the generality of mechanistic approaches.

1.3.2 Empirical Methods

Empirical models generally use correction factors that can compute the viscous performance from provided water performance of the manufacturer. These correction factors are dimensionless, and their relationship with viscous and water performance is expressed in Eqs. (8), (9) and (10).

$$Q_{\text{vis}} = C_Q Q_w \tag{8}$$

$$H_{\text{vis}} = C_H H_w \tag{9}$$

$$\eta_{\text{vis}} = C_\eta \eta_w \tag{10}$$

In Eqs. (8), (9) and (10), C_Q, C_H and C_η are the flowrate, head and efficiency correction factors, respectively. Q_w, H_w and η_w are the water flowrate, head and efficiency, respectively, extracted from the manufacturer's performance curve. Q_{vis}, H_{vis} and η_{vis} are the flowrate, head and efficiency, respectively, predicted by the correction factor-based empirical models.

Stepanoff [28] carried out early extensive experiments where he derived a chart for correction factors for the best efficiency point (BEP). It introduces a modified Reynolds number (also known as Stepanoff Reynolds number), and it assumes that for a given pump running at a constant motor speed, the head decreases as the viscosity increases in such a way that the specific speed at BEP remains constant. Stepanoff method is an iterative process that requires taking the water head and flowrate at BEP, assuming a viscous flowrate and calculating the Stepanoff Reynolds number. This number is then used to obtain the head correction factor from provided correction chart, which can be used to calculate the flowrate correction factor. The flowrate correction factor can be used to calculate the viscous flowrate and start the process again replacing the assumed viscous flowrate in the first step. The process is iterated until this viscous flowrate converges. It is to be noted that Stepanoff correction factors are only applicable at BEP.

Current industry follows standards defined by ANSI/HI 9.6.7, also known as the Hydraulic Institute method [8]. It also provides correction factor charts that take in the water head and flowrate at BEP from manufacturer's catalogue, along with the operating motor speed and fluid viscosity, to find a point simply called 'B'. This point can be used to draw a straight line that intersects with three different curves to produce the correction factors for the head, flowrate and efficiency. This method is also used at BEP, but unlike Stepanoff method, it does not require an iterative process.

Apart from Stepanoff and Hydraulic Institute methods, there have been recent developments of newer methods, such as the ones from [6, 17, 20].

Gülich [6] also proposes an empirical model that is valid for the BEP and other flowrates. This assumes that the flowrate and efficiency correction factors are independent of the operating condition, but head correction factor depends on the operating flowrate. Monte Verde [17] develops a procedure based on the model by Gülich, but proposes independent correlations for the head, flowrate and efficiency. Ofuchi et al. [20] provide with a model that can be used at BEP and other flowrates, but it lacks predictions for efficiency.

1.3.3 CFD Methods

Unlike empirical models which rely on correction factors on water test performance, CFD incorporates the fluid's properties and geometric characteristics of the pump.

A common approach to CFD uses Reynolds-averaged Navier–Stokes (RANS) with a turbulence model. Zhu et al. [34] solved the 3D steady RANS equations with standard shear-stress transport (SST) in ANSYS CFX using the 'frozen rotor' technique and a hexahedral mesh—to study the viscosity effects on the performance of an ESP. A similar study was done by Li [10] in ANSYS FLUENT. The common method is to rotate the impeller in the simulation whilst the diffuser remains fixed (known as the frozen rotor technique).

The typical work process of carrying out CFD simulations includes firstly to create or import 3D geometry of the pump. The geometry of the inlet pipe, impeller, diffuser and outlet is used to generate a fine mesh and solve for incompressible flow with appropriately set fluid properties [21, 25, 34].

Shi et al. [25] carried out CFD studies for experimental data collected for various viscosities, flowrate and motor speeds. They were successful at predicting the head within a prediction error of $\pm$ 20%.

1.4 Use of Artificial Intelligence in ESP Modelling

In recent years, artificial intelligence has been emerging as a powerful tool for complex engineering problems, including the problem of ESP behaviour under non-ideal fluids. For instance, Mohammadzaheri et al. [14] used a genetic algorithm for ESP optimisation to maximise profits. There are also studies carried out for failure predictions of ESPs by Okoro et al. [22] using neural networks and Dachanuwattana et al. [3] who used deep learning neural networks. These studies demonstrate the possibilities of applications of artificial intelligence for vastly different kinds of problems.

One of the early works on ESP head prediction with artificial intelligence was carried out by Serpa et al. [24], where they developed two different ANNs—multilayer perceptron (MLP) and neural network finite impulse response (NNFIR). Their best model outperformed an existing empirical model they chose to validate against. However, the empirical model outperformed the ANNs with validation data that were out of range of the training data of the ANNs.

Mohammadzaheri et al. [16] reviewed and replaced traditional empirical head models with an FCC neural network that was trained on experimental data collected from an ESP operating with multiphase fluids. The FCC model developed outperformed existing empirical models by a margin of 68% better MAE when evaluated on unseen testing data. In another study, with a proposed model for the same problem, Mohammadzaheri et al. [15] developed a shallow ANN. It was a simple MLP model which performed 69% better than the best empirical model in terms of MAE. These studies highlight that even simple ANNs with only a single hidden layer with five neurons, if properly trained, can capture complex multiphase fluid effects.

Ali et al. [1] investigated the performance of ESP under air–water multiphase flow using an intelligent method that utilised ANNs. They used synthetic minority over-sampling technique (SMOTE) to oversample their experimental data. A back-propagated artificial neural network (BPANN) with two hidden layers was trained and validated, resulting in a model with R^2 value of 0.988. The authors reported it to be the best model at predicting ESP performance with multiphase flow.

Sabaa et al. [23] developed four separate ANNs to predict the flowrates of oil, water, gas and liquid. The models required 16 input parameters including measurements for wellhead parameters, fluid properties and variable speed drive sensors parameters. Upon validating with testing dataset consisting of 5625 data points, the authors achieved very low MAPEs ranging from 3.7 to 6.4%. Although not related to head prediction, their work highlights the ability of ANNs at generalisation when trained correctly. Hunter et al. [7] emphasise that careful network sizing and ample training data yield better generalisation.

Beyond shallow neural networks, deep neural networks have been emerging in recent studies for ESP head predictions. Zhu et al. [32] conducted a benchmark between seven different machine learning algorithms, including deep neural networks, against mechanistic and polynomial models. Their study comprised of ESP performance under viscous and gas–liquid flows and provides an in-depth sensitivity analysis of network hyperparameters to find the optimal design of the models for each task. Key outcome of their study is that their models outperform conventional methods in terms of accuracy, practical guidelines for parameter tuning and ability to handle missing data points.

Fuzzy logic, based on fuzzy set theory, offers another method of using artificial intelligence for ESP head predictions. Mohammadzaheri et al. [13] developed a fuzzy model for head predictions of an ESP working with CO_2-diesel mixture. It is another instance where the intelligent model outperformed the best conventional method, by a 52.4% better MAE. This study indicates that a fuzzy rule with data-driven tuning can be an effective approach to ESP head prediction, where fuzzification of input variables can capture subtle effects of gas–liquid mixture that the crisp equations miss. However, there has not been any study of such methods being used for high viscosity liquids.

1.5 Problem Statement and Objectives

Conventional methods of estimating the head on an ESP using analytical or empirical approaches for non-ideal liquids—particularly that of high viscosity, suffer from limitations such as high computational cost, restrictive assumptions or narrow applicability. This demands for development of models that are quick to develop, computationally inexpensive, have better accuracy, and easier to generalise over different applications. Although [15] were able to demonstrate the use on AI for multiphase flow, there is a lack of such approach being used in ESP applications consisting of high viscosity liquids. This paper aims to address this gap by developing an AI-based model, specifically an ANN model, to predict the head of an ESP operating with different viscosities of a liquid.

The main objective of this paper is to develop two different artificial neural network (ANN) models—a multilayer perceptron (MLP) and a fully connected cascade (FCC), and evaluate their performance at predicting the head of an ESP operating with liquids of various motor speeds, flowrate and most importantly viscosities. This is done using the experimental data acquired from [25] that includes motor speeds, viscosities and flowrate which are considered as input parameters, and the head is considered as the output parameter.

The goal is to produce an efficient and accurate model that is developed using a streamlined and scalable methodology that can be used to expand the applicability of AI to a broader range of properties—beyond what is considered in this paper.

2 Methodology

2.1 Overview

This paper develops two ANNs, the specific architectures being—a multilayer perceptron (MLP) and a fully connected cascade (FCC) neural network. The choice of these architectures is motivated by their simplicity in design and prior works done by Mohammadzaheri et al. [15] and Mohammadzaheri et al. [16]. The exact design of these models has been detailed in a later section of this section.

The section discusses the acquisition and preprocessing of data, preliminary model architectures and design choices and finally tuning of the model hyperparameters.

Python is chosen as the programming language of choice for most of the workflow in this methodology. The primary reason behind this choice is that Python and all the relevant tools and libraries used for this study are well matured, free to use and are open-sourced. Not only this makes them more accessible for development, but their permissive licences also make them favourable in industrial deployments.

2.2 Data Acquisition and Preprocessing

The data used in this methodology are obtained from the experimental work of [25]. The original authors performed these experiments on a 14-stage radial-type ESP, specifically the model TE2700, with the working fluid as a lubricating oil ISO VG320. More details of their exact experimental setup and measurement techniques used can be found in their paper.

In their supplementary material, Shi et al. [25] provide four tables of experimental data, each table corresponding to a motor speed of 1800, 2400, 3000 and 3500 rpm. Each table is divided into four sections of various viscosities, and each of these sections has a column of measured flowrate and another of the head. It is to be noted that the heads provided in the tables are not the total head gained by the entire pump, it is the mean head gained per stage. A summary of the viscosities and flowrates for each motor speeds that have been measured, is provided in Table 1.

Table 1 Experimental fluid operating ranges

Motor speed (rpm)	Viscosity (cP)	Flowrate (bpd)
1800	25, 120, 240, 520	600–2100
2400	63, 107, 240, 380	800–2600
3000	25, 85, 105, 170	1400–3200
3500	90, 120, 130, 250	1600–3600

For the development of ANNs, the motor speed, viscosity and flowrate have been selected as input parameters, and the head is selected to be the output parameter. Based on these decided input and output parameters, the experimental data had to be aggregated into a single table. The table was created in Microsoft Excel with it containing four columns—the first three for the input parameters (speed, viscosity and flowrate) and the last for the output parameter (head). The table created is illustrated in Fig. 1. Note that the figure does not show all the rows of data.

The aggregated table had 531 data rows, that is, 531 different combinations of input operating conditions and their corresponding heads. This table is exported from Excel as a comma-separated value (CSV) formatted file which is a simple text-based format, making it more portable to import into other table-based applications.

	A	B	C	D	E	F
1	**Speed**	**Viscosity**	**Flowrate**	**Head**		
2	1800	120	669.04	13.52		
3	1800	120	705.1	13.28		
4	1800	120	828.35	12.81		
5	1800	120	850.3	12.29		
6	1800	120	917.16	11.86		
7	1800	120	971.88	11.54		
8	1800	120	1004.92	11.1		
9	1800	120	1026.05	10.93		
10	1800	120	1047.95	10.59		
11	1800	120	1076.4	10.5		
12	1800	120	1120.73	9.65		
13	1800	120	1269.15	8.07		
14	1800	120	1345.39	7.14		

Fig. 1 Aggregated data in excel

The aggregated data are imported into a Python notebook. The dataset is first split into three subsets—training, validation and testing. The training subset is the actual data that the ANNs train on to optimise their weights and biases, whilst the validation subset is used to prevent overfitting of the model to the training subset. The testing subset is completely unseen data that are used to evaluate the performance metrics of the developed model.

For the development of the models in this methodology, the training, validation and testing subsets are divided into ratios of 0.7, 0.15 and 0.15, respectively. This gives 371 data points for training and 80 data points each for validation and testing.

Figure 2 shows the presence of outliers in viscosity and head for the training dataset. Moreover, moderate skewness can be seen in viscosity data and a high skewness can be seen for head data as illustrated in Figs. 3 and 4, respectively. Based on these insights the training, validation and testing datasets were normalised—the speed and flowrate columns were normalised using Z-score standardisation, the viscosity column was normalised using robust scaling, and the head column was normalised using min–max normalisation. The reason for using min–max for the head column unlike robust scaling used for the viscosity column is discussed in the results section. These normalisation techniques used are non-deterministic, as such the normalisation parameters were obtained only from the training dataset and these same parameters were used to normalise the validation and testing dataset.

$$x_m' = \frac{x - \min(x)}{\max(x) - \min(x)} \tag{11}$$

The min–max normalisation technique used for the head is expressed by Eq. (11), where x and x_m' are the value and its corresponding min–max normalised value, respectively, and $\min(x)$ and $\max(x)$ are the minimum and maximum points, respectively in the given value's dataset. The selected min–max normalisation scaled the data columns into the range of [0, 1].

$$x_{z'} = \frac{x - \overline{x}}{\sigma_x} \tag{12}$$

The Z-score standardisation that is used for the speed and flowrate columns is expressed by Eq. (12), where x_z' is the z-score standardised value of x, $\overline{x}$ is the mean of x, and σ_x is the standard deviation of x. This normalisation technique produces standardised values such that the new mean is 0 and the standard deviation is 1.

$$x_{r'} = \frac{x - Q_2(x)}{Q_3(x) - Q_1(x)} \tag{13}$$

Equation (13) expresses the equation used for robust scaling. x and $x_{r'}$ are the value and its corresponding robust-scaled value, respectively. $Q_1(x)$, $Q_2(x)$ and $Q_3(x)$ are the first, second and third quartiles, respectively, of the dataset of x. This technique is used for the viscosity column to mitigate the effects of outliers and skewness. The

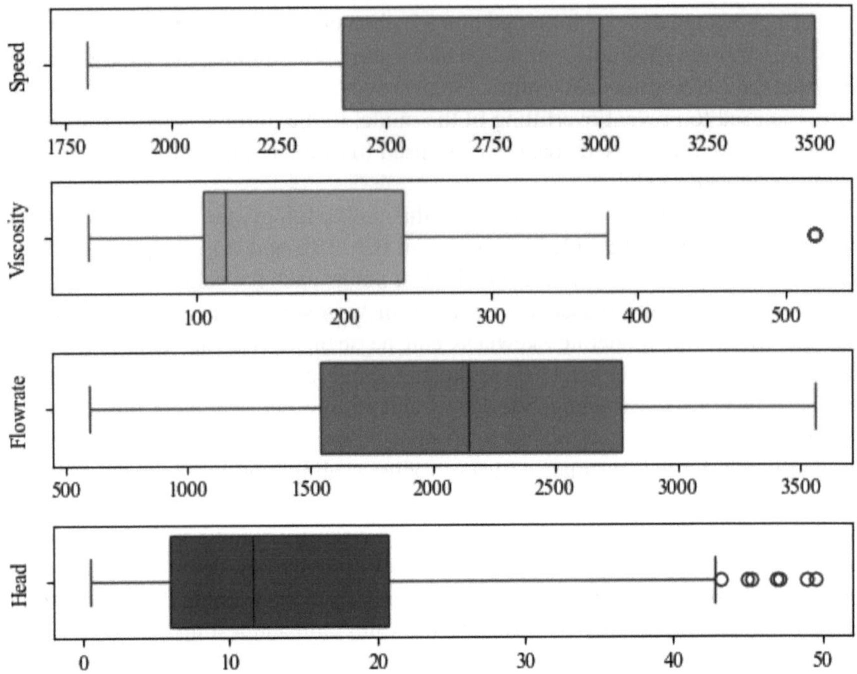

Fig. 2 Box plots for training dataset

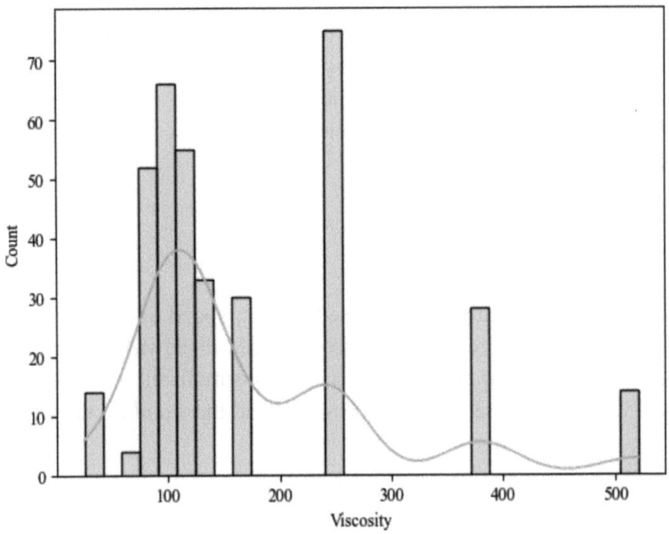

Fig. 3 Histogram of viscosity in training dataset

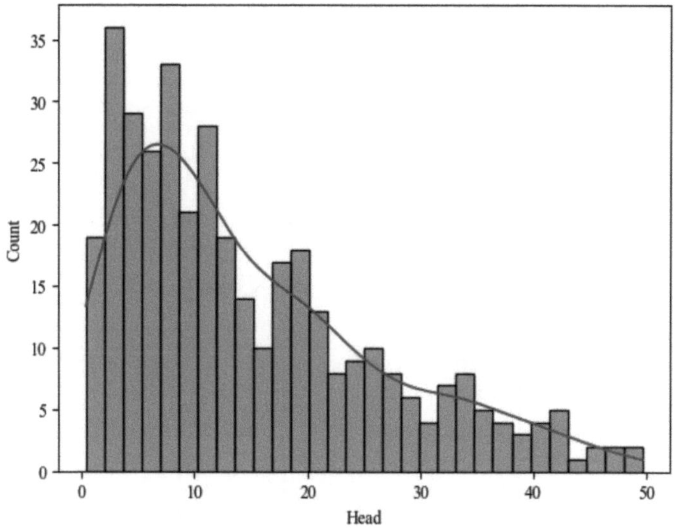

Fig. 4 Histogram of head in training dataset

new values produced are such that their median is at 0 and the scaling is determined by the interquartile range which mitigates the effect of outliers.

2.3 Preliminary Model Designs

The architectures developed for ANNs are MLP and FCC. Both the models have three units in the input layer. They both have two hidden layers with 10 neurons in each layer and a single neuron in the output layer. The choice of units in the input layer is dictated by the number of input parameters to the neural networks, that is, the speed, viscosity and flowrate. Similarly, the number of neurons in the output layer is based on the number of predicted outputs, which is just the head in this case. The number of hidden layers and their respective number of neurons are partially motivated by studies conducted by Zhu et al. [32] and Ali et al. [1], who employed deep neural networks instead of shallow ones. However, these structural choices are mostly arbitrary and are subjected to optimisation during the hyperparameter tuning step.

Figure 5 illustrates the architecture of the MLP neural network. Each unit from a given layer is connected to all the units in the subsequent layer. Each of these connections has their own weights, and each of the receiving units also has an independent associated bias.

The key difference between the MLP and the FCC neural network is that in the FCC architecture, each unit from a given layer is not only connected to the next layer but all the units in all the subsequent layers. The structure of this network is

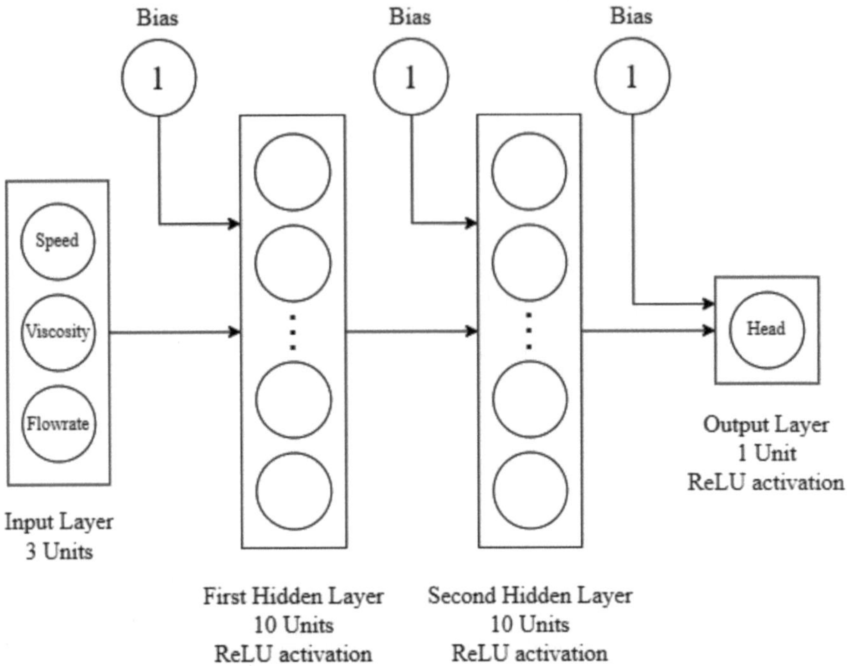

Fig. 5 MLP neural network design

illustrated in Fig. 6.

$$\text{ReLU}(s) = \begin{cases} s \text{ if } s > 0 \\ 0 \text{ if } s \leq 0 \end{cases} \tag{14}$$

The activation function for both the hidden layers and the output layer was selected to be rectified linear unit (ReLU). Equation (14) expresses the equation of ReLU activation function for a given neuron, where s is the sum of the weighted inputs to that neuron and the neuron's bias. According to Glorot et al. [5], ReLU is computationally inexpensive to calculate and achieve strong performance when used for supervised deep neural networks—perfect for architectures used in this study.

$$\text{limit} = \sqrt{\frac{6}{W_{\text{in}} + W_{\text{out}}}} \tag{15}$$

The weights of the connections in both the models were initialised using Glorot-Uniform initialiser (also known as Xavier-Uniform initialiser). It works by drawing samples from a uniform distribution within the range [−limit, limit]. The value of this limit is expressed by Eq. (15), where W_{in} and W_{out} are the number of input units and output units, respectively, in the weight tensor. The biases of all the neurons are

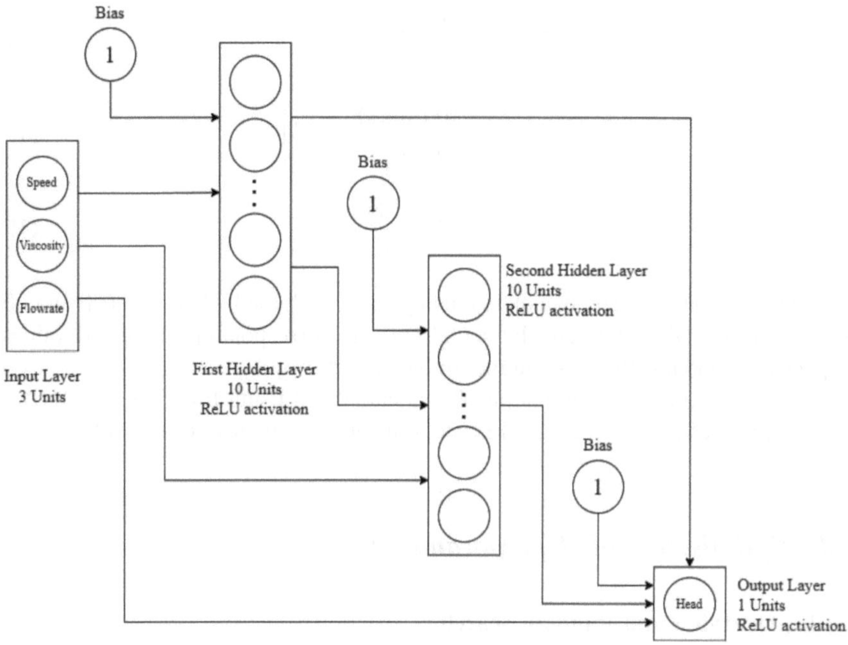

Fig. 6 FCC neural network design

initialised to zeros as proposed by Glorot et al. [5] to be coupled with the Glorot-Uniform initialiser used for the weights.

During training of the models, the optimiser used for the trainable parameters (weights and biases) was adaptive moment estimation (ADAM). This optimiser choice is based on study by Duchi et al. [4] which demonstrated ADAM to be superior optimiser for deep learning neural networks, compared to other methods.

$$\text{MSE} = \frac{1}{n}\sum_{i=1}^{n}\left(Y_i - \hat{Y}_i\right)^2 \tag{16}$$

Since the models designed solve a regression problem, the cost function (also called loss function) used for training is mean-squared error (MSE). The equation for MSE is expressed by Eq. (16), where Y_i and $\hat{Y}_i$ are the actual and the corresponding predicted values, respectively, and n is the total number of data points.

For evaluation of the developed neural networks, three kinds of metrics are used—the mean absolute error (MAE), mean absolute percentage error (MAPE) and coefficient of determination (R^2). These metrics are primarily used to evaluate against the testing subset, and sometimes against the validation subset, for example, to evaluate whether the model is overfitting.

$$\text{MAE} = \frac{\sum_{i=1}^{n} |Y_i - \hat{Y}_i|}{n} \tag{17}$$

$$\text{MAPE} = \frac{100}{n} \sum_{i=1}^{n} \left| \frac{Y_i - \hat{Y}_i}{Y_i} \right| \tag{18}$$

$$R^2 = 1 - \frac{\text{SS}_{\text{res}}}{\text{SS}_{\text{tot}}} \tag{19}$$

MAE and MAPE are calculated using Eqs. (17) and (18), respectively. In both equations, Y_i and $\hat{Y}_i$ are the actual and the corresponding predicted values, respectively, and n is the total number of data points.

The coefficient of determination (R^2) is calculated using Eq. (19), where SS_{res} and SS_{tot} are residual sum of squares and total sum of squares, respectively.

2.4 *Optimisation of Hyperparameters*

2.4.1 Overview and Random Search

The neural networks developed in previous section of the methodology are considered as preliminary models. These models are often developed to assess the problem at hand and applicability of different architectures. The models have a lot of arbitrary hyperparameters, particularly the number of hidden layers and the number of neurons in each layer, and other configurations such the training parameters optimiser, activation functions, number of epochs and so on. Manually tuning these hyperparameters makes it a tedious development process. However, this part does not need to be a manual process. The hyperparameters can optimised using various optimisation algorithms. For the optimisation process, two main things must be defined—the hyperparameter space and the objective. The hyperparameter space defines what hyperparameters can be changed and to what extent, and the objective dictates the criteria used to evaluate the models. The objective in this study is set to find the minimum validation MAE, and the defined hyperparameter space is provided in Table 2.

The hyperparameters optimisation algorithms employed in this study are Bayesian optimisation and random search. For each architecture (MLP and FCC), both optimisation algorithms are used with a maximum number of trials set to 35, and each

Table 2 Defined hyperparameter space for optimisation

Hyperparameter	Range
Number of hidden layers	1–2
Units in hidden layers	3–10
Activation function for hidden and output layers	ReLU, tanh

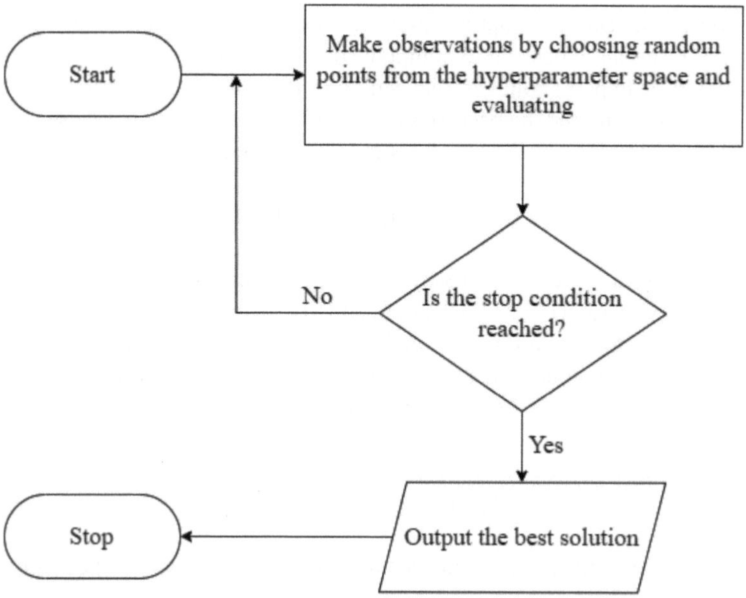

Fig. 7 Random search algorithm

trial is trained twice to mitigate randomness effect in weights initialisation. This means the entire optimisation process tries 280 neural network models (140 unique models).

The second activation function used in this optimisation is hyperbolic tangent (tanh) which is defined by Eq. (20) where *e* is Euler's number. Unlike ReLU, tanh has vanishing gradients on both ends of the independent axis which can make training deep learning networks difficult.

$$\tanh(s) = \frac{e^{2s} - 1}{e^{2s} + 1} \tag{20}$$

Random search, as the name suggests, is carried out by selecting a random combination of hyperparameters from the defined hyperparameter search space. An instance of this model is initialised with those hyperparameters, trained and evaluated against the validation subset. This is done until the stop condition is met which is in this study a defined number of maximum trials (combinations). The entire process is illustrated by a flowchart in Fig. 7.

2.4.2 Bayesian Optimisation

Unlike random search, the Bayesian optimisation algorithm used follows a very methodical approach. Like random search approach, the tuning objective is set to validation MAE. Bayesian optimisation, in this context, assumes the objective of the tuning to be an expensive-to-compute 'black box' function $f(h)$, where h is a combination of hyperparameters selected from the defined hyperparameter search space. The goal is to find the best combination of hyperparameters (h^*) such that the output of f is the global minimum. This goal is expressed in Eq. (21).

$$h^* = \arg\min f(h) \tag{21}$$

The optimisation process starts by sampling a fixed number of random combinations of hyperparameters from the defined hyperparameter space. It evaluates the function $f(h)$ for these combinations and stores the observations ($\mathcal{D}$). These observations ($\mathcal{D}$) are used to condition a probabilistic function p which is a surrogate model for f. This surrogate model is based on Gaussian process ($\mathcal{GP}$). This relationship is expressed in Eq. (22).

$$p(f|\mathcal{D}) = \mathcal{GP}\big(f, m_{f|\mathcal{D}}, K_{f|\mathcal{D}}\big) \tag{22}$$

In Eq. (22), m and K are the mean function and covariance, respectively, given the observations $\mathcal{D}$.

Using the surrogate model p,, an acquisition function is calculated. This study uses the upper confidence bound (UCB) acquisition function, defined in Eq. (23). Since the objective is to minimise f, the minimum point of the acquisition function is selected to determine the next combination of hyperparameters to be trained, evaluated and added to the observations ($\mathcal{D}$). The tuning process goes back to calculating a new surrogate model p with the new observations ($\mathcal{D}$) and continues as an iterative process until a stop condition is met (a maximum number of trials in this case).

$$a_{\text{UCB}}(h, \beta) = \mu(h) - \beta\sigma(h) \tag{23}$$

In Eq. (13), a_{UCB} is the acquisition function. $\mu(h)$ is the surrogate model's mean prediction function, and $\sigma(h)$ is the surrogate model's standard deviation function. $\mu(h)$ and $\sigma(h)$ are also known as the exploitation and exploration terms, respectively. β is a positive number called the exploration–exploitation trade-off parameter. The larger the value of β, the more likely the acquisition function is used to select an exploratory combination of hyperparameters (h)—combination where uncertainty (deviation) is higher. If the value of β is smaller, the acquisition function is more likely to choose a combination close to an already promising combination (exploitation). Note that the negative sign in Eq. (23) reflects the fact that the tuning process is aiming to minimise f.

The entire Bayesian process is summarised in the flowchart illustrated in Fig. 8.

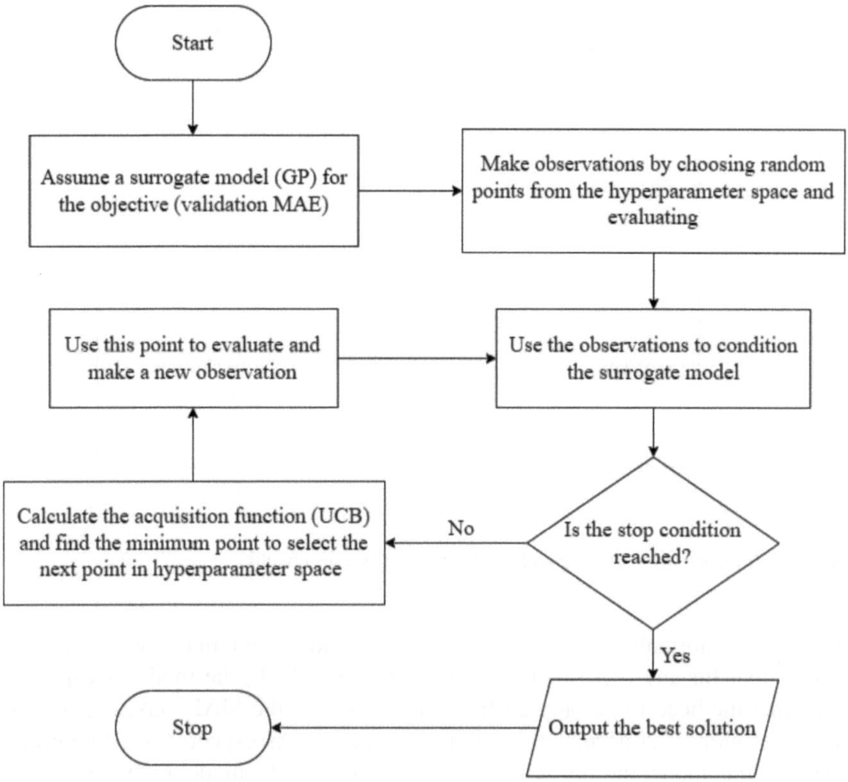

Fig. 8 Bayesian optimisation algorithm

3 Results and Discussion

3.1 Preliminary Models

One of the main steps in the methodology was to develop preliminary models to assess the problem at hand. Figure 9 shows the training and validation MAE as the MLP model trains over the training data. The model optimises the trainable parameters rather quickly within the first 50 epochs, and after the first 150 epochs, the MAE curves start to plateau. There is no sign of overfitting the model to the training data even after continuing for up to 500 epochs. The post-training validation MAE was 0.946 ft, and testing MAE was 0.996 ft—no sign of overfitting.

This preliminary MLP model is based on the normalisation described in the methodology. The methodology highlighted that the normalisation applied to the head was odd in the sense that it did not use robust scaling like the viscosity. Figure 10 illustrates the prediction performance of the MLP model with ± 15% tolerance bands. Figure 11 illustrates a similar figure, but for the MLP model trained on a dataset where

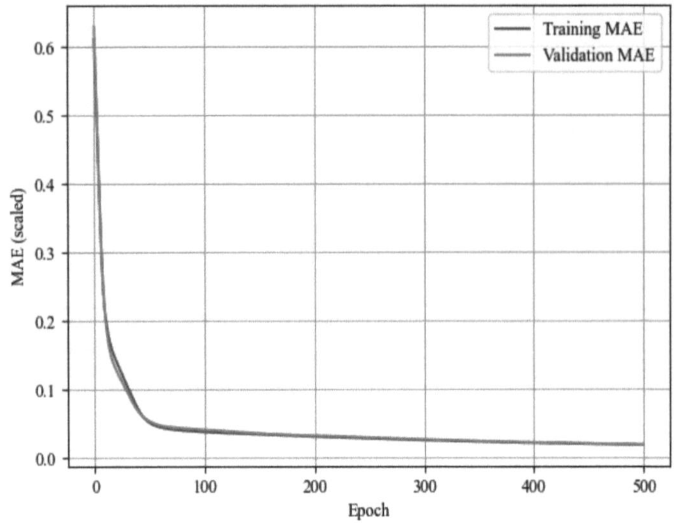

Fig. 9 Preliminary MLP training and validation MAE versus epochs

the head was normalised with robust scaling instead of min–max normalisation. It appears that for any experimental head values over 25 ft, the model is capped at predicting the head to be only 25 ft. Figure 12 shows the MAE versus epochs for the model with head normalised with robust scaling. As expected, the figure indicates that this model did not achieve MAEs as good as the model trained with head normalised with min–max normalisation—the post-training validation MAE for the robust-scaled head model was 1.5 ft.

Figure 13 illustrates the preliminary FCC model prediction performance for the validation and testing datasets with $\pm$ 15% tolerance bands. This is a comparable result to that of the preliminary MLP model. Like the MLP model, the FCC model does not show any signs of overfitting after being trained for 500 epochs. The post-training validation and testing MAE were slightly better at 0.546 and 0.601 ft, respectively. These differences are not necessarily an indicator of the FCC model being better compared to the MLP model; yet, the differences are small and can be attributed to the random nature weights initialisation. The FCC model has 204 parameters whilst the MLP has a smaller number at 161 parameters.

However, there is a significant difference between the FCC model and MLP model when trained on dataset that has the head normalised with robust scaling instead of min–max normalisation. The FCC model does not show any indication of negative effects of robust scaling like the MLP model.

Figure 14 shows the FCC model's prediction performance when trained with robust-scaled head training dataset and is clearly very different compared to MLP predictions trained on that same dataset as illustrated by Fig. 11. The post-training testing MAE of the FCC model when trained with robust-scaled head was 0.516 ft which is relatively unchanged compared to that that of when the model is trained with

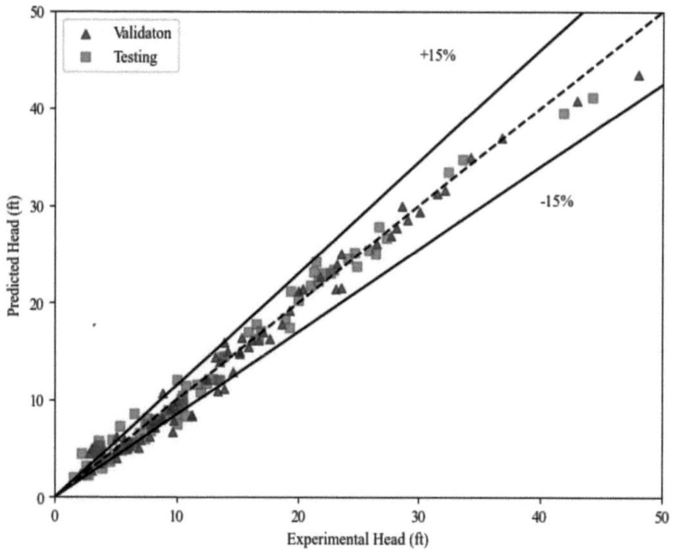

Fig. 10 Preliminary MLP prediction performance

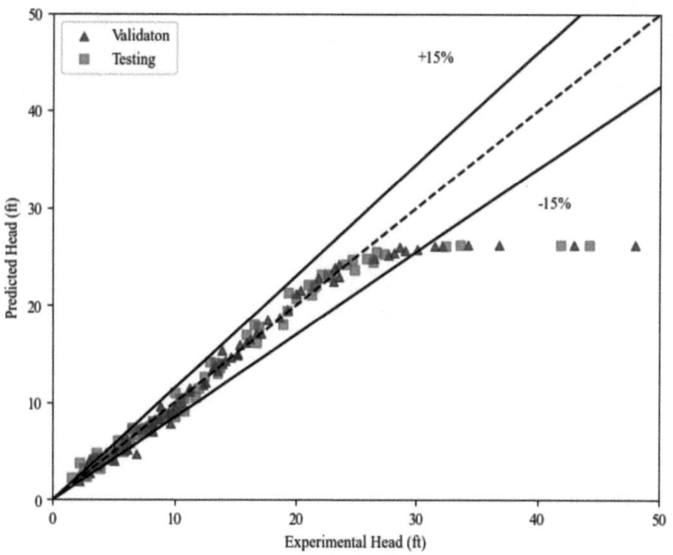

Fig. 11 Preliminary MLP predictions with robust-scaled head

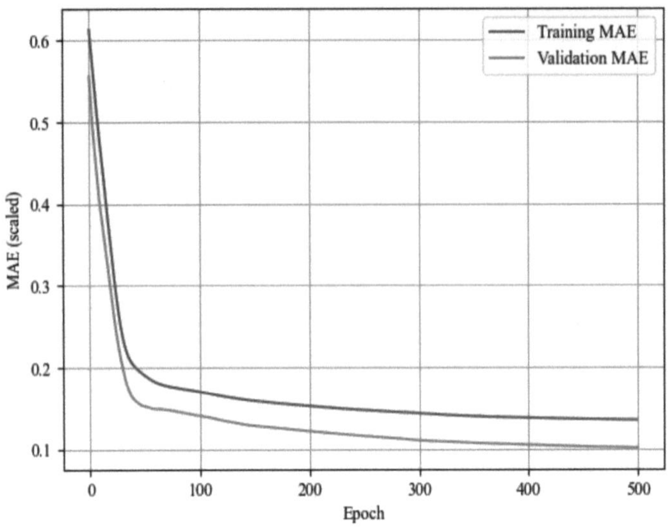

Fig. 12 Preliminary MLP MAE versus epochs with robust-scaled head

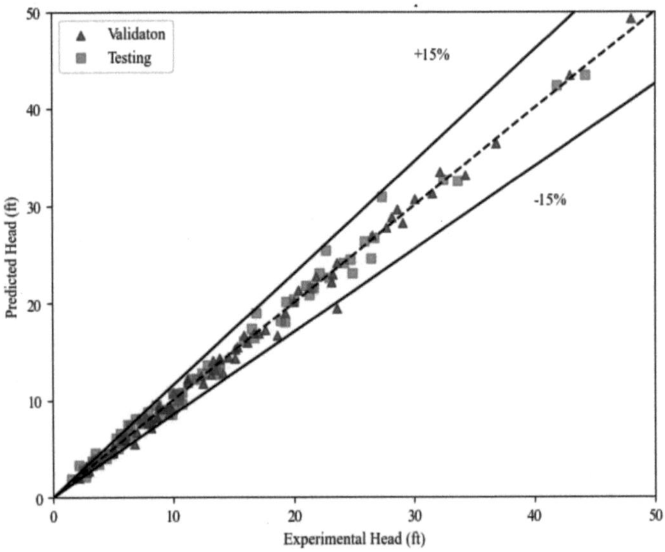

Fig. 13 Preliminary FCC prediction performance

min–max normalised head. The reasoning behind these unexpected effects of normalising the head in a certain method affecting the MLP model remains unexplored and unanswered in this study and requires further research in the future.

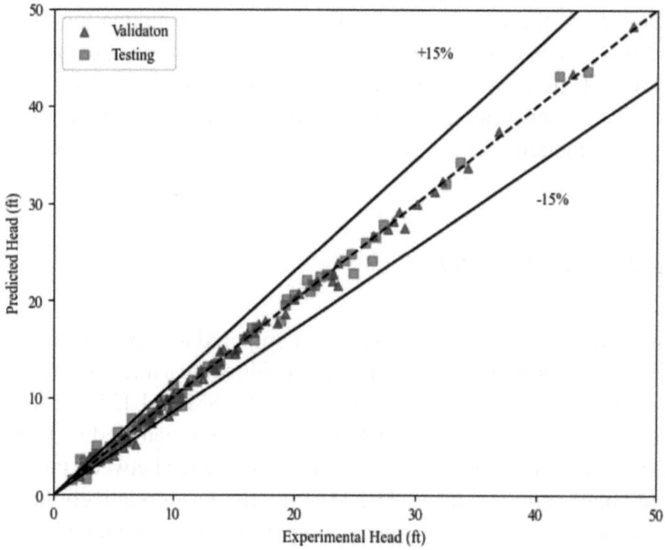

Fig. 14 Preliminary FCC predictions with robust-scaled head

3.2 *Hyperparameter Optimisation Results*

Table 3 lists the best models discovered by each tuning algorithm. Surprisingly, the best validation MAEs were discovered by random search algorithm. However, the differences in validation MAE amongst these best models are minute—they all approximate to a value of 0.012. The differences in validation MAE are therefore unprioritized, and the second criterion for the best model selection is employed—the number of model parameters. Both algorithms resulted in finding two hidden layers to be the most ideal for the best validation MAE.

For the MLP model, Bayesian optimisation produces the lower number of neurons in the hidden layers that gives a lower number of parameters. For the FCC model,

Table 3 Hyperparameter optimisation best models

Algorithm	Architecture	Hidden layer units	Hidden layer activations	Output layer activation	Validation MAE (scaled)
Random search	MLP	10 9	ReLU tanh	ReLU	0.0112141
	FCC	9 8	ReLU ReLU	ReLU	0.0122709
Bayesian optimisation	MLP	7 9	tanh tanh	ReLU	0.0119263
	FCC	10 10	tanh tanh	ReLU	0.0123763

random search finds the lower number of neurons in the hidden layers. These best models for MLP and FCC have a total of 110 and 174 parameters, respectively. These are the two models that are used in further discussions of this study.

Both the best models selected use ReLU for the activation function of the output layer, and for the hidden layers, MLP uses tanh, whereas FCC uses ReLU.

3.3 The Two Best Models

Figure 15 and 16 show the model predictions compared to experimental head for the MLP and FCC models, respectively. The graphs illustrate a tolerance band of ± 10% which is consistent to that of tolerance bands used by Shi et al. [25] for the prediction performance of their mechanistic model. Both the models can predict the head within a ± 10% error range. However, there is a slight difference between the two models, especially at small values of head. The MLP model has a relatively larger number of predictions out of the tolerance bands compared to FCC model which appears to have no more than a couple of points outside in the same range of head. The FCC model also appears to perform better at midrange head values (20–30 ft) where it predicts the data points closer to prefect prediction line (dashed line) compared to MLP model.

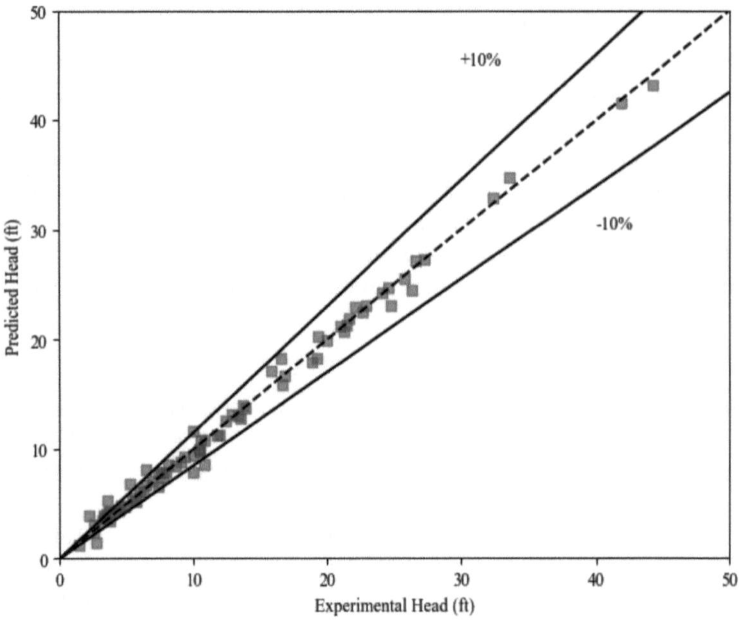

Fig. 15 Best MLP prediction performance

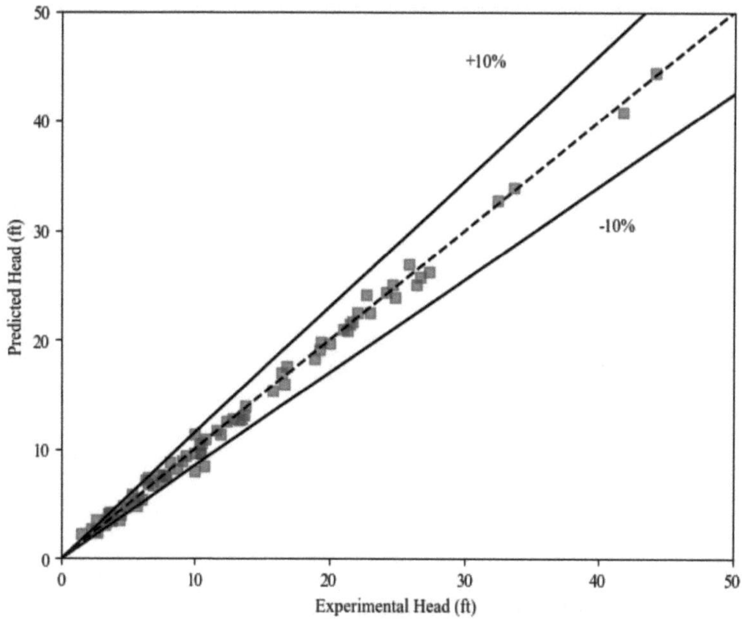

Fig. 16 Best FCC prediction performance

Figure 17 illustrates the MLP model predictions of head over changing flowrates and viscosity at motor speed of 1800 rpm. Note that the scatters include validation and testing data points, and the solid curves were drawn using all data points which includes training data. A similar illustration is created for the FCC model as shown in Fig. 18. It appears that the validation and testing datasets did not have any data points of 25 cP viscosity, as such the curve for it is missing in both the figures.

As it appears in Fig. 17, the MLP model attempts to fit a smooth curve as the flowrate changes for a given viscosity—especially for the higher viscosities; for the lower viscosities, it appears to fit a straight line. The FCC model shows a similar behaviour for the lower viscosities, but for higher viscosities, Fig. 18 shows that the model fits curves or lines with abrupt gradient changes as the flowrate changes.

The tendency of FCC model to produce these abruptly changing head curves becomes more prominent with data points of higher motor speeds. Figure 19 illustrates the head curves produced by FCC model at motor speed of 3000 rpm. The FCC model clearly shows a tendency to draw out a line through all the points instead of attempting to draw out a single smooth curve. However, the MLP model draws out a single smooth curve for each viscosity as illustrated in Fig. 20. This would suggest that the FCC model is more susceptible to noise in experimental measurements of individual data points instead of 'looking at the big picture'. The MLP model instead produces smooth curves which fit traditional intuition and expectations of what the head curve must look like, especially considering the head curve of water.

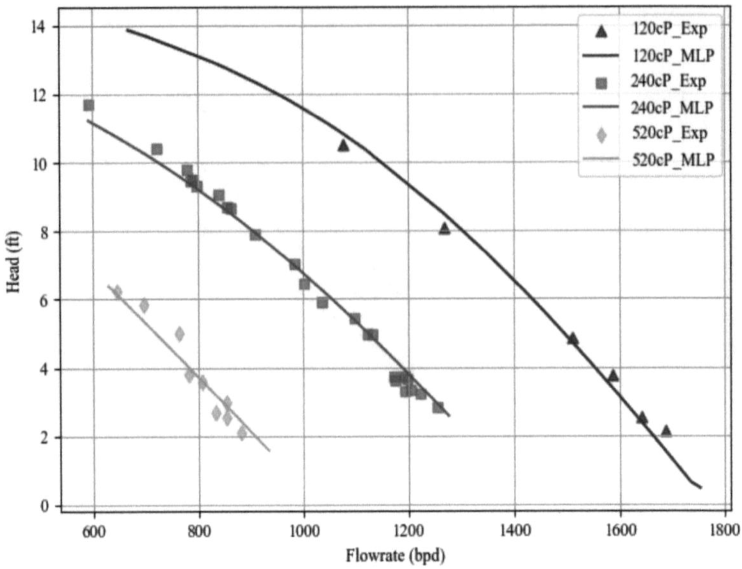

Fig. 17 MLP predicted head curve at 1800 rpm

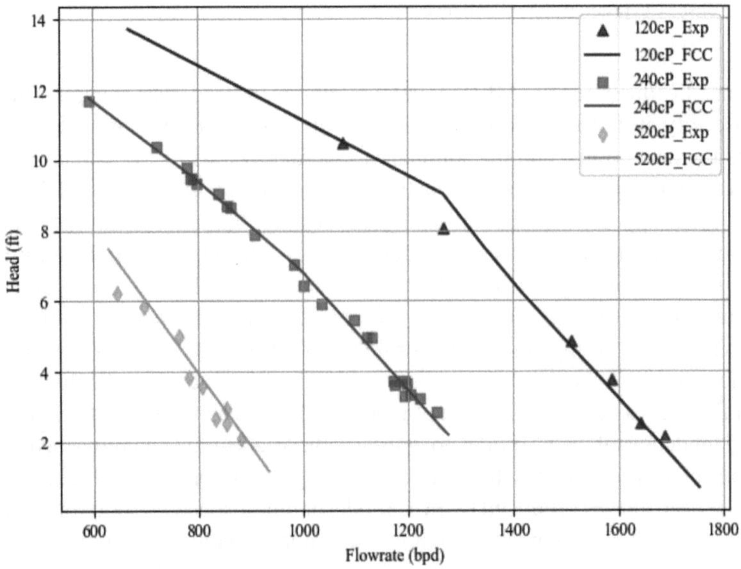

Fig. 18 FCC predicted head curve at 1800 rpm

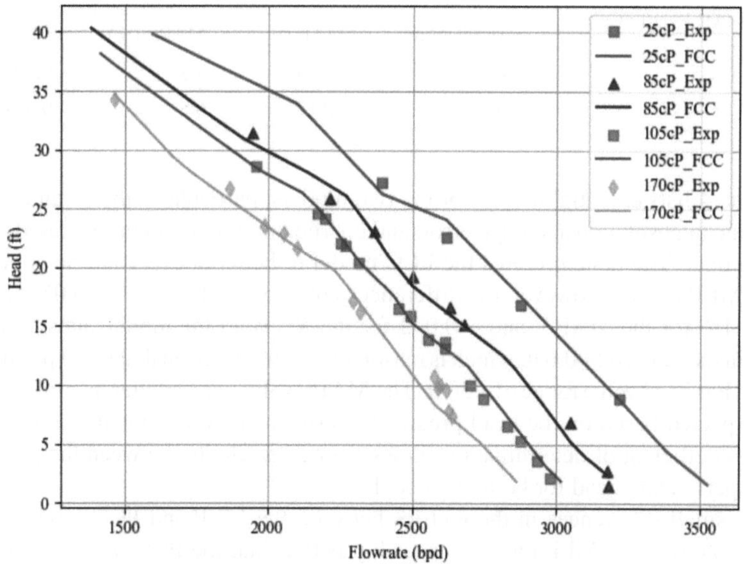

Fig. 19 FCC predicted head curve at 3000 rpm

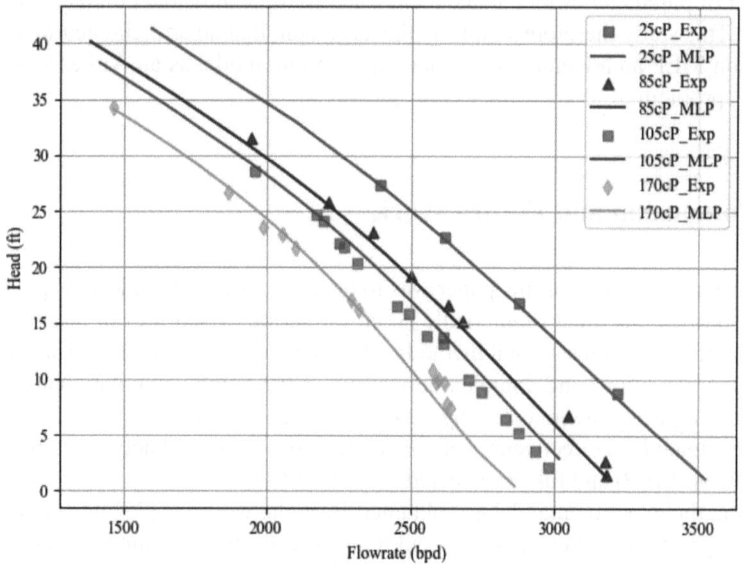

Fig. 20 MLP predicted head curve at 3000 rpm

Table 4 MLP and FCC metrics

Architecture	MAE (ft)	MAPE (%)	R^2
MLP	0.569	7.8	0.9928
FCC	0.512	6.8	0.9949

Table 4 lists essential metrics of the two models, particularly the mean absolute error, mean absolute percentage error, and coefficient of determination based on the testing data. The table presents the FCC model to better at every metric compared to the MLP model. However, the differences are very small, that is, ~ 0.06 ft for the MAE, 1% for the MAPE, and ~ 0.002 for the R^2. Both the models have a testing MAE in the range 0.5–0.6 ft, which is reasonably accurate given that the experimental dataset has head value range of ~ 50 ft. The MAPE values are smaller than 10% which was expected based on the head prediction performances illustrated in Figs. 7 and 8. The coefficient of determination shows strong correlation between the predicted and experimental head for both the models.

The small differences in the metrics between the MLP and FCC model makes it arguable that the MLP model is possibly better than the FCC model, especially considering the head prediction curves where the MLP model behaves more intuitively correct and given the fact that the MLP model requires significantly smaller number of parameters which make them faster to train and more efficient in deployments. This study therefore stands by this argument that an artificial neural network based on MLP architecture is the more appropriate model as an AI-based solution for this problem.

4 Conclusion and Future Work

The primary objective of this paper was to develop AI-based models for predicting the head of an electrical submersible pump for various viscosities of liquids. Given the limitations of conventional methods—like analytical methods requiring detailed operating conditions and complex derivations, CFD being computationally expensive and empirical methods having narrow applicability of correction factors—this study evaluated the effectiveness of two architectures of neural networks: multilayer perceptron (MLP) and fully connected cascade (FCC).

Using the experimental dataset obtained from the works of [25] that contained a wide range of viscosities, flowrates and motor speeds, both the models were trained and validated using a well-structured methodology. The methodology consisted of data aggregation, data preprocessing with insights-driven normalisation, development of preliminary models to assess the problem and finally application of random search and Bayesian optimisation to tune the hyperparameters of the preliminary models.

The results demonstrated that both the MLP and FCC models, after hyperparameter tuning, achieved high predictive accuracy, with testing mean absolute errors

(MAEs) under 0.6 ft and mean absolute percentage errors (MAPEs) under 8%. Even though the FCC model seemed to outperform the MLP model based on raw metrics, the MLP model exhibited a more consistent and intuitive head prediction curve that align more closely to traditionally expected head curve. Moreover, the MLP model required fewer parameters, making it faster to train and potentially more suitable for real-time or embedded applications.

In conclusion, this study successfully demonstrated that simple AI-based approaches—particularly the MLP architecture, can provide accurate, efficient and scalable solutions to ESP head estimation compared to conventional methods. These models may be modified to account for broader applications such as ESP configuration and other fluid properties.

However, certain observations made in the results regarding the unexpected sensitivity of the MLP model to the output parameter normalisation remain unexplored and unexplained in this paper, requiring further investigation. Future work may focus on understanding internal mechanisms of MLP architectures driving this behaviour. Another focus of future studies could be to extend and incorporate additional input parameters such as the pump geometry which could generalise the models over different types of pumps. Moreover, the hyperparameter tuning process may be enhanced by adopting a new optimisation objective that considers both validation MAE and model complexity, such as the number of trainable parameters. Introducing such a multi-objective cost function during model tuning could lead to architectures that are not only accurate but also lightweight and efficient—making them better suited for real-time deployment or use in resource-constrained environments.

Appendix 1: System Information

All the code executed in this study were done in a Google Colab Python notebook. The environment and system information are provided as follows:

- Operating system: Ubuntu 24.04.4 LTS
- GCC version: 11.4.0
- Python version: 3.11.13
- Tensorflow version: 2.18.0
- Keras version: 3.8.0
- Keras-tuner version: 1.4.7
- CPU: Intel(R) Xeon(R) CPU @ 2.20GHz
- RAM: 13,341,992 kB

Appendix 2: Data Preprocessing

The dataset was split and normalised using the following block of Python code in the next page:

```
import pandas as pd

from sklearn.model_selection import train_test_split
from sklearn.preprocessing import StandardScaler, RobustScaler,
MinMaxScaler
from sklearn.compose import ColumnTransformer

df = pd.read_csv("dataset.csv")
rnd_seed = 1746448054

input_transformer = ColumnTransformer(transformers=[
    ('std', StandardScaler(), ["Speed", "Flowrate"]),
    ('robust', RobustScaler(), ["Viscosity"])
])
output_transformer = MinMaxScaler()

input_columns = ["Speed", "Viscosity", "Flowrate"]
output_columns = ["Head"]

training_df, buffer_df = train_test_split(df, test_size=0.3,
random_state=rnd_seed)
validation_df, testing_df = train_test_split(buffer_df, test_size=0.5,
random_state=rnd_seed)

training_input_df, training_output_df = training_df[input_columns],
training_df[output_columns]
validation_input_df, validation_output_df =
validation_df[input_columns], validation_df[output_columns]
testing_input_df, testing_output_df = testing_df[input_columns],
testing_df[output_columns]

training_input_df_norm =
pd.DataFrame(input_transformer.fit_transform(training_input_df),
columns=input_columns)
training_output_df_norm =
pd.DataFrame(output_transformer.fit_transform(training_output_df),
columns=output_columns)

validation_input_df_norm =
pd.DataFrame(input_transformer.transform(validation_input_df),
columns=input_columns)
validation_output_df_norm =
pd.DataFrame(output_transformer.transform(validation_output_df),
columns=output_columns)

testing_input_df_norm =
pd.DataFrame(input_transformer.transform(testing_input_df),
columns=input_columns)
```

Appendix 3: Preliminary Model

```
"""
The preliminary MLP model is created using the following block of Py-
thon code
"""

import tensorflow as tf

model = tf.keras.Sequential([
  tf.keras.Input((3,), name="layers_input"),
  tf.keras.layers.Dense(10, activation="tanh", name="layers_hidden_1"),
  tf.keras.layers.Dense(10, activation="relu", name="layers_hidden_2"),
  tf.keras.layers.Dense(1, activation="tanh", name="layers_output")
], name="MLP")

model.compile(optimizer="adam", loss="mse", metrics=["mae"])
```

```
"""
The preliminary FCC model is created using the following block of Py-
thon code
"""

import tensorflow as tf

input_layer = tf.keras.layers.Input(shape=(3,), name="layers_input")
hidden_layer_1 = tf.keras.layers.Dense(10, activation="relu",
name="layers_hidden_1")(input_layer)
hidden_layer_2_concat = tf.keras.layers.Concatenate()([input_layer,
hidden_layer_1])
hidden_layer_2 = tf.keras.layers.Dense(10, activation="relu",
name="layers_hidden_2")(hidden_layer_2_concat)
output_layer_concat = tf.keras.layers.Concatenate()([input_layer,
hidden_layer_1, hidden_layer_2])
output_layer = tf.keras.layers.Dense(1,
name="ouput_layer")(output_layer_concat)

model = tf.keras.models.Model(inputs=input_layer, outputs=output_layer,
name="FCC")

model.compile(optimizer="adam", loss="mse", metrics=["mae"])
```

Appendix 4: Hyperparameter Tuners

Refer to the following pages.

```
"""
The following block of code is used to generate MLP models from the
given hyperparameter space
"""

import tensorflow as tf

from keras_tuner import HyperParameters
from keras_tuner.tuners import RandomSearch, BayesianOptimization


def build_model_mlp(hp: HyperParameters):
  model = tf.keras.Sequential()
  model.add(tf.keras.layers.Input(shape=(3,)))

  for num_layer in range(hp.Int("num_layers", 1, 2)):
    model.add(tf.keras.layers.Dense(
        units=hp.Int(f"units_{num_layer}", min_value=3, max_value=10,
step=1),
        activation=hp.Choice(f"activation_{num_layer}", ["relu",
"tanh"]),
        name=f"layers_hidden_{num_layer}"
    ))

  model.add(tf.keras.layers.Dense(1,
activation=hp.Choice(f"activation_output", ["relu", "tanh"])))

  model.compile(optimizer="adam", loss="mse", metrics=["mae"])

  return model
```

```
"""
The following block of code is used to generate FCC models from the
given hyperparameter space
"""

import tensorflow as tf

from keras_tuner import HyperParameters
from keras_tuner.tuners import RandomSearch, BayesianOptimization


def build_model_fcc(hp: HyperParameters):
  inputs = tf.keras.layers.Input(shape=(3,))

  layers = []

  for num_layer in range(hp.Int("num_layers", 1, 2)):
    units = hp.Int(f"units_{num_layer}", min_value=3, max_value=10,
step=1)
    activation = hp.Choice(f"activation_{num_layer}", ["relu", "tanh"])

    if layers:
      concat = tf.keras.layers.Concatenate()(layers + [inputs])
    else:
      concat = inputs

    dense = tf.keras.layers.Dense(units, activation=activation)(concat)
    layers.append(dense)

  final_concat = tf.keras.layers.Concatenate()(layers + [inputs])
  output = tf.keras.layers.Dense(1,
activation=hp.Choice(f"activation_output", ["relu",
"tanh"]))(final_concat)

  model = tf.keras.models.Model(inputs=inputs, outputs=output)
  model.compile(optimizer="adam", loss="mse", metrics=["mae"])

  return model
```

```
# This is used to configure the tuners and run the tuning process

tuner_mlp_rand = RandomSearch(
    build_model_mlp,
    objective="val_mae",
    max_trials=35,
    executions_per_trial=2,
    directory="model_tunings",
    project_name="mlp_tunings_rand"
)

tuner_mlp_bayes = BayesianOptimization(
    build_model_mlp,
    objective="val_mae",
    max_trials=35,
    executions_per_trial=2,
    directory="model_tunings",
    project_name="mlp_tunings_bayes"
)

tuner_fcc_rand = RandomSearch(
    build_model_fcc,
    objective="val_mae",
    max_trials=35,
    executions_per_trial=2,
    directory="model_tunings",
    project_name="fcc_tunings_rand"
)

tuner_fcc_bayes = BayesianOptimization(
    build_model_fcc,
    objective="val_mae",
    max_trials=35,
    executions_per_trial=2,
    directory="model_tunings",
    project_name="fcc_tunings_bayes"
)

tuner_mlp_rand.search(training_dataset, validation_data=validation_da-
taset, epochs=400, verbose=0)
tuner_mlp_bayes.search(training_dataset, validation_data=validation_da-
taset, epochs=400, verbose=0)

tuner_fcc_rand.search(training_dataset,
validation_data=validation_dataset, epochs=400, verbose=0)
tuner_fcc_bayes.search(training_dataset,
validation_data=validation_dataset, epochs=400, verbose=0)
```

References

1. Ali A, Yuan J, Iqbal S, Si Q, Raza Shah Naqvi SM, Hussain S, Shah F, seif Ghith E, Saleem K (2025) An intelligent computing methodology for two-phase flow performance assessment of electrical submersible pump using artificial neural network and synthetic minority oversampling technique. Measurement 244:116512. https://doi.org/10.1016/j.measurement.2024.116512
2. Çengel YA, Cimbala JM (2014) Affinity laws. In: Fluid mechanics (3rd ed). McGraw-Hill
3. Dachanuwattana S, Ratanatanyong S, Wasanapradit T, Vimolsubsin P, Kulchanyavivat S (2022) The deployment of deep learning models for performance optimization and failure prevention of electric submersible pumps. Day 2 Wed. https://doi.org/10.4043/31612-MS
4. Duchi J, Hazan E, Singer Y (2011) Adaptive sub gradient methods for online learning and stochastic optimization. J Mach Learn Res 12:2121–2159. http://jmlr.org/papers/v12/duchi11a.html
5. Glorot X, Bordes A, Bengio Y (2011) Deep sparse rectifier neural networks. In: Proceedings of the fourteenth international conference on artificial intelligence and statistics, pp 315–323. https://proceedings.mlr.press/v15/glorot11a/glorot11a.pdf
6. Gülich JF (2010) Centrifugal pumps. Springer Berlin Heidelberg.https://doi.org/10.1007/978-3-642-12824-0
7. Hunter D, Yu H, Pukish IMS, Kolbusz J, Wilamowski BM (2012) Selection of proper neural network sizes and architectures—a comparative study. IEEE Trans Industr Inf 8(2):228–240. https://doi.org/10.1109/TII.2012.2187914
8. Hydraulic Institute (1955) Determination of pump performance when handling viscous liquid (10th ed)
9. Lea JF, Rowlan L (2019) Electrical submersible pumps. In: Gas well deliquification. Elsevier, pp 237–308. https://doi.org/10.1016/B978-0-12-815897-5.00012-3
10. Li W-G (2008) Numerical study on behavior of a centrifugal pump when delivering viscous oils-part 1: performance. Int J Turbo Jet Eng 25(2). https://doi.org/10.1515/TJJ.2008.25.2.61
11. Li W-G (2013) Effects of flow rate and viscosity on slip factor of centrifugal pump handling viscous oils. Int J Rotating Mach 2013:1–12. https://doi.org/10.1155/2013/317473
12. Li W-G (2014) Mechanism for onset of sudden-rising head effect in centrifugal pump when handling viscous oils. J Fluids Eng 136(7). https://doi.org/10.1115/1.4026882
13. Mohammadzaheri M, AlQallaf A, Ghodsi M, Ziaiefar H (2018) Development of a fuzzy model to estimate the head of gaseous petroleum fluids driven by electrical submersible pumps. Fuzzy Inf Eng 10(1):99–106. https://doi.org/10.1080/16168658.2018.1509523
14. Mohammadzaheri M, Tafreshi R, Khan Z, Franchek M, Grigoriadis K (2016) An intelligent approach to optimize multiphase subsea oil fields lifted by electrical submersible pumps. J Comput Sci 15:50–59. https://doi.org/10.1016/j.jocs.2015.10.009
15. Mohammadzaheri M, Tafreshi R, Khan Z, Ghodsi M, Franchek M, Grigoriadis K (2020) Modelling of petroleum multiphase flow in electrical submersible pumps with shallow artificial neural networks. Ships Offshore Struct 15(2):174–183. https://doi.org/10.1080/17445302.2019.1605959
16. Mohammadzaheri M, Tafreshi R, Khan Z, Ziaiefar H, Ghodsi M, Franchek M, Grigoriadis K (2019) Modelling of electrical submersible pumps for petroleum multiphase fluids, an intelligent approach supported by a critical review and experimental results. J Eng Res [TJER] 16(2):77. https://doi.org/10.24200/tjer.vol16iss2pp77-86
17. Monte Verde W (2016) Modelagem do desempenho de bombas de BCS operando com misturas gás-óleo viscoso [Universidade Estadual de Campinas]. https://doi.org/10.47749/T/UNICAMP.2016.976346
18. Monte Verde W, Kindermann E, Biazussi JL, Estevam V, Foresti BP, Bannwart AC (2023) Experimental investigation of the effects of fluid viscosity on electrical submersible pumps performance. SPE Prod Oper 38(01):1–19. https://doi.org/10.2118/210603-PA
19. Nguyen T (2020) Electrical submersible pump, pp 107–179. https://doi.org/10.1007/978-3-030-40720-9_3

20. Ofuchi EM, Cubas JMC, Stel H, Dunaiski R, Vieira TS, Morales REM (2020) A new model to predict the head degradation of centrifugal pumps handling highly viscous flows. J Petrol Sci Eng 187:106737. https://doi.org/10.1016/j.petrol.2019.106737
21. Ofuchi EM, Stel H, Sirino T, Vieira TS, Ponce FJ, Chiva S, Morales REM (2017) Numerical investigation of the effect of viscosity in a multistage electric submersible pump. Eng Appl Comput Fluid Mech 11(1):258–272. https://doi.org/10.1080/19942060.2017.1279079
22. Okoro EE, Sanni SE, Okigbo A, Adeyemi F, Emetere ME, Obomanu T (2021) Predicting non-newtonian fluid electric submersible pump failure using deep learning and artificial neural network. Proc Inst Mech Eng, Part A: J Power Energy 655(1):012027. https://doi.org/10.1088/1755-1315/655/1/012027
23. Sabaa A, Abu El Ela M, El-Banbi AH, Sayyouh MHM (2023) Artificial neural network model to predict production rate of electrical submersible pump wells. SPE Prod Oper 38(01):63–72. https://doi.org/10.2118/212284-PA
24. Serpa A, Sassim N, Biazussi JL, Monte Verde W, Barrera Salamanca LF (2017) Use of artificial neural networks for identification of the electric submersible pump performance working with a viscous fluid. In: Procceedings of the 24th ABCM international congress of mechanical engineering. https://doi.org/10.26678/ABCM.COBEM2017.COB17-0304
25. Shi Y, Zhu J, Wang H, Zhu H, Zhang J, Zhang H-Q (2021) Experiments and mechanistic modeling of viscosity effect on a multistage ESP performance under viscous fluid flow. Proc Inst Mech Eng, Part A: J Power Energy 235(8):1976–1991. https://doi.org/10.1177/09576509211014974
26. Sirino T, Stel H, Morales REM (2013) Numerical study of the influence of viscosity on the performance of an electrical submersible pumphttps://doi.org/10.1115/FEDSM2013-16374
27. Stel H, Sirino T, Ponce FJ, Chiva S, Morales REM (2015) Numerical investigation of the flow in a multistage electric submersible pump. J Petrol Sci Eng 136:41–54. https://doi.org/10.1016/j.petrol.2015.10.038
28. Stepanoff AJ (1957) Centrifugal and axial flow pumps (2nd ed). Wiley
29. Sun D, Prado M (2006) Single-phase model for electric submersible pump (ESP) head performance. SPE J 11(01):80–88. https://doi.org/10.2118/80925-PA
30. Takacs G (2009) ESP components and their operational features. In: Electrical submersible pumps manual. Elsevier, pp 51–118. https://doi.org/10.1016/B978-1-85617-557-9.00003-8
31. Torabi R, Nourbakhsh SA (2016) The effect of viscosity on performance of a low specific speed centrifugal pump. Int J Rotating Mach 2016:1–9. https://doi.org/10.1155/2016/3878357
32. Zhu H, Yu H, Sun Q, Wang Q, Jing H, Abdikadyrov R (2025) Deep learning-based performance prediction of electric submersible pumps under viscous and gas-liquid flow conditions. Machines 13(2):135. https://doi.org/10.3390/machines13020135
33. Zhu H, Zhu J, Zhang H-Q (2022) Mechanistic modeling of gas effect on multi-stage electrical submersible pump (ESP) performance with experimental validation. Chem Eng Sci 252:117288. https://doi.org/10.1016/j.ces.2021.117288
34. Zhu J, Banjar H, Xia Z, Zhang H-Q (2016) CFD simulation and experimental study of oil viscosity effect on multi-stage electrical submersible pump (ESP) performance. J Petrol Sci Eng 146:735–745. https://doi.org/10.1016/j.petrol.2016.07.033
35. Zhu J, Zhang H-Q (2018) A review of experiments and modeling of gas-liquid flow in electrical submersible pumps. Energies 11(1):180. https://doi.org/10.3390/en11010180
36. Zhu J, Zhu H, Cao G, Zhang J, Peng J, Banjar H, Zhang H-Q (2020) A new mechanistic model to predict boosting pressure of electrical submersible pumps under high-viscosity fluid flow with validations by experimental data. SPE J 25(02):744–758. https://doi.org/10.2118/194384-PA

Fault Diagnosis of a Vehicle Cylinder Head: An AI Approach

Injamamul Haque, Morteza Mohammadzaheri, Payam Soltani, and Loay Al-Lawati

Abstract The research applies data-driven methods to locate faults in automotive cylinder heads by analyzing vibrations with supervised machine learning and signal processing techniques. The main goal focuses on determining structural fault positions by analyzing acceleration data through frequency domain features. The Fast Fourier Transform (FFT) processed time-series signals which originated from finite element simulation of controlled impact tests. The transformation process revealed amplitude spectra which allowed researchers to identify the most significant frequency components. The variance-based feature selection method identified five frequency bins which showed the highest variability between faulty cases to detect unique fault location characteristics. The features were used as inputs to create two predictive models which included a linear regression model with Least Squares Estimation (LSE) and a nonlinear Multi-Layer Perceptron (MLP) neural network. The models received their training and testing data from a predefined dataset split to maintain consistent evaluation results. The MLP model performed better than the LSE model according to the results. The MLP achieved a Mean Absolute Error (MAE) of 1.362 mm, whereas the LSE resulted in an MAE of 8.098 mm, both over a 270 mm fault range. The research shows that nonlinear models outperform linear approaches because they better detect complex vibration patterns. The MLP model achieved stable training convergence which confirms its reliable performance. The research confirms that FFT signal processing combined with machine learning algorithms produces precise fault detection results. The proposed method provides a non-destructive and expandable system for automotive component health monitoring which requires additional validation to transition to real-world diagnostic applications.

Keywords Fault diagnosis · Cylinder head · Vibration analysis · Fast Fourier transform · Multi-layer perceptron · Least-squared estimation

I. Haque · M. Mohammadzaheri (✉) · L. Al-Lawati
German University of Technology in Oman, Athaibah, Sultanate of Oman
e-mail: morteza.zaheri@gutech.edu.om

P. Soltani
Military Technological College, Muscat, Oman

A. Azizi (ed.), *Smart Diagnostics and Predictive Maintenance*, Emerging Trends in Mechatronics, https://doi.org/10.1007/978-981-95-3921-5_4

Abbreviations

AI	Artificial intelligence
ANN	Artificial neural network
DFT	Discrete Fourier transform
DIT	Decimation in time
ESP	Electric submersible pump
FFT	Fast Fourier transform
FEA	Finite element analysis
FEM	Finite element model
LSE	Least squares estimation
MAE	Mean absolute error
ML	Machine learning
MLP	Multi-layer perceptron
MSE	Mean squared error
PCA	Principal component analysis
RQ	Research question
SFFS	Sequential forward floating selection
SVM	Support vector machine
MRA	Multiresolution analysis

List of Symbols

X	Input feature matrix (27 samples × 5 features)
y	Output vector of fault positions (target values in mm)
$\hat{h}$	Estimated weights in linear regression using LSE
h	Number of hidden neurons in the MLP
s	Number of input features (5 in this study)
Δf	FFT frequency resolution
T	Total duration of the signal (0.02 s)
Var	Sample variance across all faulty test cases for a frequency bin
x_i	FFT amplitude at a given frequency bin for the i-th sample
$\overline{x}$	Mean FFT amplitude across all faulty samples for a given bin
n	Number of samples (27 in this case)
MAE	Mean absolute error
y_i	True output (fault location) for the i-th sample
$\hat{y}_i$	Predicted output for the i-th sample
$\sum$	Summation symbol (used in variance and MAE equations)
$\lvert\cdot\rvert$	Absolute value operator (used in MAE)
X^{T}	Transpose of the input matrix X

1 Introduction

1.1 Background

The complex engine system depends on reliable operation of essential components under extreme pressure and temperature conditions. The cylinder head serves as a vital component which protects combustion chambers while containing intake and exhaust ports together with coolant channels. The detection of cylinder head cracks remains a serious issue because it results in coolant leakage and compression loss that may lead to total engine destruction when not properly identified [1]. Research findings indicate that fatigue cracking occurs most commonly in the bridge region which connects intake and exhaust valve ports because of thermal stress and mechanical forces [2]. Early identification of such faults becomes essential to prevent engine damage while maintaining safety standards.

Traditional inspection methods for detecting cylinder head faults involve time-consuming disassembly operations followed by visual examination or testing the cooling system under pressure. Traditional inspection methods prove to be time-consuming while detecting minor cracks only after failure occurs. The need for a diagnostic method without engine disassembly has motivated interest in vibration-based detection systems. The method of vibration analysis remains a leading condition monitoring technique for predictive maintenance because vibrational signals contain extensive information about machinery health [3]. The introduction of cracks in structural components such as cylinder heads leads to modifications in their dynamic behavior which results in changes to their vibrational responses during excitation. Research on vibration-based damage detection methods has been extensive to detect changes in dynamic behavior [4]. Damage detection through analysis of frequency spectra and mode shapes makes it possible to identify small damage before it reaches critical dimensions. The Fast Fourier Transform (FFT) provides a proven method for fault diagnosis through frequency domain analysis because it detects fault-specific "signatures" within rotating machinery and structural components.

Artificial intelligence together with machine learning has brought new advanced capabilities to fault diagnosis during the last few years. Expert interpretation together with baseline comparisons of vibrational features remain the standard requirement for traditional vibration analysis techniques. Data-driven AI algorithms possess the capability to discover intricate patterns within sensor data for automatic diagnostic operations. Artificial neural networks have demonstrated successful performance when they identify structural damage through analysis of vibration characteristics. The study of Hakim et al. [5] demonstrated how artificial neural networks can accurately detect both the location and measure the extent of damage in beam-like structures through modal vibration feature analysis. Research demonstrates that AI can execute fault detection missions which prove challenging for analytical models to accomplish. Automotive engine fault detection research has started to employ machine learning techniques by scientific investigators. A vibration-based diagnosis system for automobile cylinder heads was developed by Taajobian et al. [6] through exciting the part and recording its acceleration response where Fourier analysis extracted features for Multi-Layer Perceptron (MLP) neural network training to detect faults along with their locations. The research revealed that short vibration tests lasting about half a second along with low-frequency signals below 1.2 kHz provided high accuracy in detecting faults in cylinder heads [6]. Vibration-based feature extraction when combined with AI-driven classification presents a promising method for successful fault diagnosis.

Research in modern fault diagnosis depends on simulation tools for producing diagnostic data. Engine physical experiments become too costly or impractical for use since intentionally creating cylinder head faults for vibration data collection remains impossible in most laboratory settings. FEA enables virtual testing through the creation of a detailed 3D cylinder head model which allows researchers to introduce faults for studying their impact on structural dynamics. ANSYS together with FEA software enables users to simulate how a cylinder head vibrates under excitation conditions both when the component is intact and when it experiences damage. Simulation models have proven capable of replacing experimental studies for creating diagnostic algorithms when validated according to Taajobian et al. [6]. The simulation method enables the systematic generation of vibrational response data for multiple fault conditions before signal processing techniques like FFT are used to identify diagnostic features that separate faulty from healthy conditions. The obtained features serve as input to train and evaluate AI models. This methodology develops a digital duplicate of the component for diagnostic functions which follows the direction of current intelligent predictive maintenance systems.

The research background emerges from the convergence between structural dynamics and signal processing and artificial intelligence. AI enables automated detection of cylinder head fault changes through analysis of its vibration signature. The following sections will define the specific problem addressed, the motivation for tackling it, and the scope of the research, before outlining the objectives.

1.2 Problem Statement

Engine design improvements have not solved the difficult task of identifying faults in essential components such as the cylinder head. The detection of cylinder head faults through teardown inspection remains challenging because the initial symptoms (e.g., engine vibration changes, coolant leakage, exhaust gases in coolant) appear only when the fault reaches an advanced stage. The main issue addressed in this chapter involves the absence of a quick non-destructive method to identify structural problems in automobile cylinder heads. The practice of scheduled inspections followed by waiting for obvious signs of failure results in unexpected engine failures and expensive maintenance costs. The individual potential of vibration-based methods and AI-driven diagnostics needs integration to create a practical system for detecting cylinder head faults. The research focuses on identifying cylinder head faults through the analysis of vibrational data by employing machine learning models.

The main challenge lies in accurately identifying structural faults through cylinder head vibration signatures. The supervised learning approach transforms vibrational features into fault positions while using simulations to produce data which solves an essential deficiency in current automotive maintenance procedures by allowing fault detection before engine disassembly.

1.3 Motivation and Relevance

The research aims to achieve practical and scientific objectives through this investigation. Automotive reliability and safety benefits directly from improved engine component fault diagnosis methods. The presence of an unmonitored cylinder head fault can cause engine overheating or total engine failure that could abandon passengers and cost thousands of dollars for repairs. The detection of faults at early stages enables maintenance teams to replace faulty components before minor problems develop into major issues. Such predictive maintenance leads to both lower costs and increased safety for vehicle owners together with fleet operators. The vibration test studied in this chapter can become part of regular maintenance procedures or onboard diagnostic systems since it serves as an early warning system for engine crack detection like modern sensors do for different vehicle faults. Predictive maintenance shows significant economic value because proactive fault detection produces better results than reactive failure management. The scientific value of this research extends to the broader structural health monitoring and machine learning applications within engineering fields. The research adds to the ongoing development of AI-based approaches that replace conventional engineering diagnostic methods. The implementation of AI models specifically neural networks along with mechanical engineering knowledge to interpret vibrational data represents an interdisciplinary advancement. AI model training data comes from FEA simulations which serves as an innovative approach for the future. The digital model of mechanical components allows researchers to create and validate detection methods within controlled environments before physical application becomes possible. Such monitoring systems become more important because of the Industry 4.0 era. The research has significant relevance because it follows current scientific developments as recent studies show AI-based fault detection systems improve both automation and sensitivity for rotating machinery and engines according to Tama et al. [3]. The research addresses essential maintenance cost reduction and engine failure prevention in automotive applications while developing advanced diagnostic methods through simulation and signal processing and artificial intelligence.

1.4 Research Questions

Based on the problem statement and objectives of this study, the following key research questions (RQs) are asked:

RQ1: How can vibrational analysis be utilized to detect a crack in an automobile cylinder head? In particular, what changes in the vibration signal (e.g., in the frequency domain) and does it indicate to us that there is fault present in the head?

RQ2: What are the most informative features of the cylinder head's vibration response for fault diagnosis? Specifically, can the dominant frequencies extracted from the vibration signal via FFT reliably distinguish between a healthy cylinder head and a faulty one?

RQ3: How effective are machine learning models in diagnosing a cylinder head fault from vibration derived features? This includes sub-questions such as: (a) Can a simple linear regression model achieve acceptable accuracy in detecting the fault? And (b) and does a more complex model like a Multi-Layer Perceptron (neural network) improve the fault detection performance compared to the linear model?

These research questions help guide the investigation by focusing on the detection mechanism (vibration analysis), the data features to be used (frequency components), and the chosen diagnostic approach (comparing the linear model with an MLP). Answering these questions will demonstrate the feasibility and benefit of the proposed AI-based fault diagnosis method for cylinder heads.

1.5 Scope and Limitations

The scope of this chapter is defined by several deliberate constraints. Firstly, the study considers only a single type of fault–a spherical fault in the cylinder head. Other engine faults (such as valve defects, gasket leaks, or misfires) are outside the scope. The fault is assumed to be an identifiable structural discontinuity (e.g., a through-thickness spherical fault at a critical location). Secondly, all data used in this research are generated via simulation. The vibrational responses of the cylinder head, both healthy and faulty, are obtained from Finite Element Analysis in ANSYS. While this approach provides controlled and repeatable data, it means that real engine operating conditions (e.g., combustion vibrations, noise from other components) are not captured. This is a limitation, as the simulation may not account for all complexities of an actual engine environment. The frequency range analyzed is limited by the simulation setup and the interest in relatively low-frequency vibrations ~2 kHz (in line with prior findings that frequencies below ~1.2 kHz can suffice for detecting such faults [6]. Very high-frequency phenomena or acoustic emission are not considered.

In terms of the diagnostic approach, the scope is confined to vibration signal processing and supervised learning models. The features are extracted using FFT, and specifically the analysis in this chapter focuses on a handful of dominant frequency peaks from the spectrum (the five most dominant frequencies). Other advanced feature extraction techniques (e.g., wavelet transforms or time–frequency methods) are beyond the scope. Similarly, the machine learning models explored are limited to two types: a linear model and a Multi-layer Perceptron (neural network). The linear model provides a baseline (representing a simple classifier, such as linear regression or logistic regression used for fault identification), while the MLP represents a more flexible nonlinear model. No other classifiers (such as support vector machines, decision trees, or deep convolutional networks) are implemented. This is a limitation, as it leaves out potentially better-performing algorithms; however, the intention here is to demonstrate the concept rather than extensively optimize the fault locating method.

Another limitation pertains to fault characterization. The chapter primarily aims to detect the presence of a fault (fault location). Estimation of fault size is not deeply explored and remains limited. The simulation assumes a certain fault geometry based on the scenario of interest; variations in fault size or multiple concurrent faults are not analyzed, which could limit the generality of the results. Finally, the validation of the diagnostic method is done on simulated data. While the simulation is designed to be realistic (and draws on material properties and boundary conditions reflective of a real engine cylinder head), the true test of any diagnostic method would be on actual physical measurements. Due to time and resource constraints, such experimental validation lies outside the scope of this work. The results and conclusions must therefore be interpreted with the understanding of this simulation-to-reality gap.

1.6 Objectives

The primary objective of this chapter is to develop and evaluate an AI-based method for diagnosing a fault in an automobile cylinder head using vibrational analysis. This overall goal can be broken down into the following specific objectives.

Simulate Faulty and Healthy Conditions: Create a detailed finite element model of an automobile cylinder head using ANSYS, and simulate its dynamic response under excitation for two scenarios: (a) an intact (healthy) cylinder head, and (b) the same cylinder head with a simulated fault. The simulation should provide time domain vibration signals (e.g., acceleration versus time) that reflect how the structure responds in each condition.

Extract Vibration Signal Features: Process the simulated vibration signals to extract meaningful features that can differentiate between healthy and faulty conditions. In particular, apply the Fast Fourier Transform to convert each time domain signal into a frequency spectrum, and identify the important frequency components. From each spectrum, select a set of features specifically, the five dominant frequency peaks (and their amplitudes) which will serve as the input features for the fault diagnosis models that are used.

Develop and Train Diagnostic Models: Implement two diagnostic models in MATLAB, one using a linear regression approach and one using a Multi-Layer Perceptron (neural network). Train these models on the simulated data so that they learn to recognize the patterns associated with a fault. The linear model provides a baseline for performance, while the MLP is expected to capture more complex relationships between the frequency features and the presence of a fault.

Evaluate Model Performance: Assess the accuracy and reliability of the trained models in detecting the cylinder head fault. This involves testing the models on data not seen during training (or a partition of the simulation data reserved for testing). The objective is to determine whether the models can correctly distinguish the location of the fault on the cylinder head cases based on vibration features, and then compare the performance of the linear model with the MLP using Mean Absolute Error (MAE).

Analyze and Discuss Results: Although technically part of the chapter output rather than a separate objective, an important final step is to analyze the results in the context of the research questions. This includes interpreting if the frequency features that were selected were most indicative of the fault location, and drawing conclusions about the viability of the proposed fault diagnosis approach.

By accomplishing these objectives, the chapter will demonstrate a proof of concept framework for fault diagnosis of a cylinder head using AI and comparing it with linear model and compare them. This framework encompasses the full pipeline from simulation and signal processing to model training and validation.

To provide a foundation for the proposed research, the following sections review existing literature on fault diagnosis in automotive engine components, emphasizing vibration-based methods and AI-driven approaches. This review synthesizes key studies, highlights their significance, identifies gaps, and establishes the context for developing an AI-based diagnostic framework for cylinder head faults.

1.7 Background

1.7.1 Importance of Fault Diagnosis in Automotive Engine Components

Cylinder heads are essential components of internal combustion engines, forming the upper part of the combustion chamber and supporting important functionalities such as airflow, fuel injection, and coolant regulation. Their structural integrity is vital for maintaining compression and efficient engine operation. However, these components are regularly subjected to extreme thermal cycles and mechanical stresses, making them prone to fatigue cracking, casting defects, and impact-related failures. If left unnoticed, structural faults in the cylinder head can disrupt engine functions, resulting in coolant escaping, combustion instability, and eventually mechanical breakdown [7].

A parallel can be drawn with pipeline systems, where internal pressure and cyclic stresses often lead to fatigue cracks, highlighting the urgency of timely fault detection in similarly stressed components like engine heads. As with pipelines, timely detection in engines can prevent catastrophic failures and significant maintenance costs. The recent research demonstrates that analyzing shifts in dynamic characteristics of the cylinder head can reveal early stage damage through vibration signatures [6]. The concept is grounded in the understanding that faults whether they be cracks, voids, or loosened components—alter the stiffness and mass distribution of the structure, which in turn modifies its vibrational response.

Moreover, vibration signals captured from the cylinder head have been successfully used to diagnose faults such as valve clearance irregularities and injector malfunctions [8, 9]. These signals carry information about combustion stability, fuel injection timing, and valve impacts, all of which shift when a fault is present. As Barelli et al. [10] noted, vibration and acoustic measurements are non-intrusive yet highly correlated with in-cylinder pressure behavior, making them ideal for continuous engine monitoring.

In summary, effective fault diagnosis in engine components, especially the cylinder head, is critical for maintaining performance, reducing maintenance downtime, and ensuring operational safety. Vibration analysis emerges as a promising non-destructive approach that captures early changes in dynamic properties indicative of the internal structural issues.

1.7.2 Traditional Methods for Diagnosing Cylinder Head Cracks

Historically engine cylinder head crack detection required traditional methods which included visual examination combined with pressure testing and coolant analysis and magnetic particle inspection. Visual inspection represents the simplest method that technicians perform after disassembly. Visual examination reveals both surface damage and discolorations yet it fails to identify hidden problems or internal defects [11]. The pressure testing procedure that uses compressed air or fluid to detect leaks can find coolant passage-related cracks yet it misses microcracks and unrelated flaws [12]. Coolant analysis detects chemical contamination such as hydrocarbons and metals in engine coolant but this method lacks precision in both fault location and severity assessment [13].

The use of these conventional methods in detection comes with multiple limitations. The inspection process requires extensive manual work and complete engine dismantling for proper evaluation which leads to increased downtime and higher labor expenses. Their detection capability for internal and developing defects remains weak because these methods usually fail to detect cracks at their initial stages. The study by Krstic et al. [13] on aircraft engine cylinder head failures demonstrated that standard inspections failed to detect internal flaws or casting porosity which resulted in catastrophic fatigue failure before discovery. Post failure analysis revealed that, Xu and Yu [12] found metallurgical flaws such as graphite clustering that standard inspections failed to detect.

Modern industrial diagnostic techniques are gaining acceptance because traditional methods have proven ineffective for their intended purposes. Engineers can analyze internal component states through vibration-based analysis and FFT-driven signal processing which eliminate the requirement for component disassembly. The modern diagnostic methods provide superior early crack detection capabilities which lead to predictive maintenance instead of traditional reactive repair practices [14]. The value of high sensitivity non-destructive techniques becomes more evident in industrial engine diagnostics as the limitations of traditional testing methods become apparent.

1.7.3 Shift Toward Vibration-Based Fault Diagnosis

Vibration-based diagnostics serve as a strong non-destructive testing alternative to traditional fault detection methods especially for internal engine components like cylinder heads because of their evident performance limitations. The procedure includes generating external impulses by using instrumented hammers to acquire vibration measurements from the system [6]. Internal flaws such as microcracks and casting anomalies produce distinct changes in dynamic characteristics which manifest as shifts in natural frequencies and amplitude spectrum variations [15]. The high sensitivity of dynamic properties to structural integrity makes vibration signature analysis an effective method for detecting invisible defects.

The main advantage of vibration-based diagnosis exists because it functions well in actual industrial settings. Chu et al. [16] showed that vibration analysis can detect structural abnormalities in operational components. Their experimental results demonstrated that this detection method can find early stage faults without requiring disassembly or system interruption thus making it suitable for critical time-dependent systems like automotive engines.

According to Burda et al. [17], vibration signals from rotating systems and load bearing elements provide essential information regarding mechanical system condition. Their review showed that real-time monitoring combined with advanced signal processing including band pass filtering and envelope detection enhances diagnosis in high noise environments. The method becomes expandable through these improvements for multiple applications within mechanical domains. The precision of vibration measurements has enhanced because of recent developments in sensor technology and data acquisition methods. Modern high-resolution accelerometers together with higher sampling rates enable engineers to detect minor damage through precise frequency response measurement. The advancement in hardware equipment supports frequency domain analysis to create a method which is both practical and scalable for predictive maintenance.

The method provides a swift and budget-friendly and non-contact solution for structural health monitoring. The method uses internal defects to modify vibrational behavior which enables both fault localization and proactive maintenance planning [15–17].

1.8 Significance of the Research

1.8.1 Why Vibration + FFT is Advantageous

Fast Fourier Transform (FFT) has become a standard method for converting vibration signals into the frequency domain because it reveals hidden system dynamics which time domain signals often conceal. The frequency domain representation of data enables better detection of periodic components that reveal structural anomalies such as cracks or defects because raw temporal data contains noise and complex waveform interactions. The detection of mechanical degradation in rotating machinery and engine components depends heavily on frequency response analysis because degradation produces distinct patterns in frequency response [18].

The computational efficiency of FFT becomes more pronounced through Decimation In Time (DIT) algorithm optimization which allows fast processing of large datasets according to Xiang et al. [18]. Compared to time domain analysis, FFT simplifies the feature extraction process, which in turn supports faster model training and diagnostic decision making. The feature extraction process becomes easier with FFT which leads to faster model training and diagnostic decision making compared to time domain analysis. The application of FFT features in mechanical fault classification achieves high accuracy according to the recent studies which support both machine learning and deep learning models [19].

The primary factor affecting FFT analysis accuracy is the choice of resolution. The practice of zero padding for resolution enhancement produces artificial frequency content that leads to misleading peak identification [20]. The use of actual signal duration for natural resolution maintains the authentic frequency characteristics of the system. The recent literature emphasizes that this approach generates features which are both physically interpretable and highly reliable for diagnosis [18, 19]. This study uses natural FFT resolution to maintain features which correspond to actual cylinder head vibrational behavior.

1.8.2 Importance of Fault Signatures and Variance

In the context of vibration-based diagnostics, a fault signature refers to the identifiable change in a signal most commonly in the frequency domain that results from structural defects such as cracks/faults, loosened components, surface wear or any faults. For mechanical components like cylinder heads or rolling bearings, these signatures are derived by comparing the FFT-transformed vibration data of a faulty sample to that of a healthy reference. The difference in amplitude at various frequencies forms a pattern that is characteristic of the fault, which can then be utilized as an input for intelligent diagnostic systems [6].

The concept of variance plays a critical role in feature selection within such datasets. As vibration signals inherently contain thousands of frequency bins post FFT, not all of them are equally informative. Variance quantifies how much each frequency fluctuates across different tests, highlighting instability tied to faults across multiple samples at each frequency point. Frequencies with higher variance are typically more sensitive to changes caused by faults and thus serve as more robust discriminators in supervised learning models [21]. Using variance-based selection allows engineers to reduce high-dimensional FFT data to a compact set of the most informative frequencies, which in turn improves model training efficiency and minimizes the risk of overfitting.

Jayaswal et al. [22] highlighted that fault signatures in frequency spectra often manifest as unique amplitude modulations and sidebands, depending on the type and location of the defect. These patterns, however, are embedded in noisy and complex signal environments, especially under real world operating conditions. Therefore, isolating the most sensitive frequencies via statistical variance enhances the signal to noise ratio of the diagnostic process and enables earlier and accurate fault detection.

Furthermore, Ayman et al. [23] emphasized that feature learning is arguably the most challenging phase in machine condition monitoring. They observed that shallow methods like statistical variance, though manually driven, are still widely employed as a reliable first stage filter especially when integrated into hybrid pipelines with deep learning. In such frameworks, variance-based filtering not only accelerates convergence but also helps avoid irrelevant or redundant frequency features that may otherwise confuse the learning algorithm.

As a result, the adoption of fault signature extraction followed by variance-based selection enables efficient, interpretable, and computationally lightweight modeling making it highly suitable for real time applications in automotive and rotating machinery diagnostics.

1.8.3 Why AI Helps in Fault Localization

Artificial intelligence (AI) now serve as a powerful method to detect faults in mechanical systems through their ability to extract subtle nonlinear patterns from noisy vibration data in cylinder heads. Linear models like Least Squares Estimation (LSE) work well for straightforward connections between variables but fail to extract complex relationships between input signals and their responses. The flexibility of Multi-Layer Perceptrons (MLPs) enables them to learn nonlinear relationships between features (e.g., FFT amplitudes) and fault locations, and this paper will prove this statement comparing both linear and nonlinear models evaluating their performance. The research by Tama et al. [3], demonstrates that these models successfully process high-dimensional vibration signal features better than linear classifiers when detecting early stage or partially masked defects through their hierarchical representations.

The generalization capability of deep learning models proves effective even when operating in environments with artifacts. Fuchs et al. [24] showed that convolutional networks along with 3D U-Nets can detect flaws with high accuracy in real-world scans after training on simulated CT data. Their research shows that models trained with sufficient variations learn to disregard domain-specific noise or imaging artifacts which can be applied to vibration diagnostics for recognizing faults despite sensor noise and structural resonances. AI technology provides increased value to applications operating in dynamically changing operating conditions particularly when machinery speeds change. Sowmya et al. [25] demonstrate that traditional methods fail during non-stationary operations because frequency features become indistinct while transients appear. AI systems leverage signal processing methods such as generalized demodulation, order tracking and multi-sensor time–frequency fusion to develop fault patterns that adapt to changing time conditions. The approaches improve resistance to speed changes and enable AI systems to maintain their accuracy during different operational modes. The study demonstrates that LSE and MLP models show different performance results when evaluated through Mean Absolute Error (MAE) across 27 distinct test cases. The integration of AI not only improves predictive performance but also opens the door to automated, scalable fault localization systems that can operate under variable speed, noisy environments, and even in the presence of simulation trained inputs something previously thought infeasible.

1.8.4 Practical Impact and Application

The integration of vibration-based diagnostics into maintenance workflows presents substantial practical benefits for industrial systems, particularly in automotive engine testing. Traditional maintenance approaches such as scheduled servicing or running to failure strategies are associated with increased operational costs and unanticipated downtime. In contrast, vibration analysis supports condition-based maintenance by providing early indicators of internal faults, allowing timely intervention and optimized equipment usage [26].

By converting time domain data into frequency domain profiles through techniques like FFT, engineers can isolate patterns indicative of specific faults such as bearing wear, imbalance, or looseness without dismantling the machinery. This not only accelerates fault localization but also enables the deployment of automated testing routines where manual inspections are impractical [16]. As highlighted in Chu et al. [16], vibration analysis is especially valuable in detecting faults within systems that are otherwise inaccessible or embedded, such as Electric Submersible Pumps (ESP) or engine heads during operation.

Moreover, the practical impact of combining vibration diagnostics with AI-driven fault recognition is reflected in the increasing trend toward predictive analytics. According to Mohd Ghazali and Rahiman [27], the growing adoption of real time monitoring tools and smart sensors enables fault detection without halting operations. This improves decision making, reduces unscheduled maintenance, and extends component lifespan. Their review underscores that over 82% of fault detection applications in rotating machinery now incorporate vibration analysis due to its diagnostic precision.

As a result, the proposed methodology leveraging utilizing signals, FFT, and intelligent modeling offers a scalable, low cost, and effective solution for early fault detection in engine cylinder heads, directly contributing to reduced testing time, lower maintenance costs, and improved system reliability.

1.9 *Review of Key Studies on Fault Diagnosis in Mechanical Systems*

1.9.1 FFT and Frequency Domain Analysis in Vibration Studies

Vibration-based diagnostics depends heavily on the Fast Fourier Transform (FFT) as a fundamental technique for analyzing rotating machinery. FFT serves as a basic transformation tool which converts time domain signals into frequency domain signals to detect mechanical faults through their specific frequency components. A DSP-based FFT analyzer showed its value in motor monitoring according to Betta et al. [28] through real-time FFT analysis which detects balance defects and gear issues by analyzing harmonic patterns and sidebands. The research included fault emulation through frequency domain manipulation to provide model-based diagnosis with high accuracy during dynamic operating conditions.

The accuracy and resolution of frequency depend on how the FFT algorithm is executed. Luo et al. [20] criticized the use of zero padding because it improves spectral interpolation but does not enhance signal interpretability. Luo et al. [20] developed an interpolated DFT method which works with various windowing approaches yet recognized that systematic errors occur in noisy signals and near Nyquist frequency conditions. The FFT algorithm with natural resolution maintains its accuracy to machine behavior without using zero padding.

The diagnostic capabilities of FFT heavily rely on accurate models for both healthy and faulty system states even though it operates well under stationary conditions. A DSP-based analyzer according to Betta et al. [28] demonstrated real-time FFT processing abilities for asynchronous motors. The system distinguished mechanical faults through harmonic and tone analysis while utilizing a waveform generator to create synthetic fault conditions for model-based diagnostics. The implementation demonstrated how FFT identifies distinct spectral changes due to unbalance and misalignment and gear defects through proper analysis of frequency domain data with reference spectra. Luo et al. [20] pointed out that zero padding enhances visual frequency resolution yet produces no actual improvement because it only fills gaps between spectral bins and the sampling duration controls the real resolution. Luo et al. [20] developed a powerful interpolation method that minimizes such artifacts especially when the signal contains noise. According to Al-Badour et al. [29], FFT-based methods show insufficient performance when dealing with non-stationary signals which appear during quick machine startups and brief rubbing events. The fast evolution of signal characteristics exceeds FFT's ability to track so researchers use time–frequency methods including wavelets for these situations. FFT serves as an essential computation-efficient diagnostic core for stationary vibrations and well-windowed inputs because it continues to be vital for many diagnostic systems.

The research conducted by Al-Badour et al. [29] demonstrated how FFT compares to wavelet-based methods. The stationary condition performance of FFT works well but it fails to detect transient characteristics and signal discontinuities that occur during rubbing or startup transient events. The essential use of FFT for fault signature identification in periodic signals continues even though it needs improvement for handling transient events through hybrid methods. The research proves FFT serves as a dependable diagnostic technique when researchers select appropriate resolution limits while effectively controlling spectral artifacts.

1.9.2 Statistical Feature Selection for Vibration Signals

The FFT processing of vibration signals generates hundreds to thousands of frequency bins. The abundant data becomes problematic because it creates high-dimensional data that leads to overfitting and computational inefficiency in diagnostic models. The statistical feature selection method decreases data dimensions to prevent overfitting and maintain diagnostic performance. The most popular method for selecting frequency components uses variance analysis because it detects the most varying components between conditions which are considered fault-sensitive [6].

The research by Ghaderi and Kabiri [30] used FFT to segment acoustic emission data from automobile engines followed by Correlation-based Feature Selection (CFS). The method selected features that strongly correlated with class distinctions while reducing the overlap between selected variables which resulted in enhanced classification accuracy across multiple frequency resolutions. The CFS method picked features mainly from high-frequency segments because these components demonstrated high sensitivity to fault patterns despite environmental noise.

The research conducted by De Assis Boldt et al. [31] showed that efficient feature selection plays a crucial role in rotating machinery diagnostics. The hybrid method which combined ranking and wrapper approaches decreased the number of features by more than 50% without compromising classification accuracy. The authors demonstrated that wrapper methods produce better diagnostic results but their execution speed improves when using intelligent pre-ranking techniques for practical real-time applications.

The MRA FFT-based pipeline for induction motor fault detection in real time was validated by Agustin et al. [32] through feature selection. The Correlation and Fitness Function-based Selection (CFFS) method achieved 88% testing accuracy under ideal noise conditions by finding a balance between model accuracy and feature parsimony.

The research demonstrates that fault diagnosis systems require compact statistically informative features to develop effective lightweight systems.

1.9.3 ML Models for Fault Detection (Linear Versus MLP)

Machine learning models use vibration-based fault diagnostics to transform signal features into particular fault conditions. The combination of interpretability and low computational cost exists in traditional linear models through Least Squares Estimation (LSE). The fundamental limitation of linear models exists because they fail to detect the nonlinear relationships which appear in vibration signals when system behavior changes due to dynamic faults and noise. The domain benefits from nonlinear models like Multi-Layer Perceptrons (MLPs) because these models excel at learning complex mappings between high-dimensional inputs and fault labels.

The authors of Ghate and Dudul [33] developed an optimized MLP classifier for induction motor fault detection through statistical analysis of current signal features. The research achieved 100% classification accuracy across different conditions by using PCA for dimensionality reduction which also protected the system from noisy environments. The authors of Goyal et al. [34] conducted a performance evaluation between Support Vector Machines (SVMs) and Artificial Neural Networks (ANNs) for non-contact bearing diagnosis. The authors determined that both models reached equivalent classification performance after PCA dimensionality reduction and SFFS selection but the ANN model showed superior adaptability to changing load and speed conditions. Liyanagedera et al. [35] further developed this research by training a backpropagation neural network with Fourier transformed vibration signals from internal combustion engines. The model achieved superior results in fault combination detection including misalignments and imbalance because it used 256 input nodes and 150 hidden neurons to process nonlinear input–output relationships.

The research confirms that MLPs excel at vibration-based fault diagnosis when dealing with high-dimensional and nonlinearly separable input features [33–35].

1.9.4 Studies Using Fast Fourier Transform and Artificial Intelligence

New research shows how Fast Fourier Transform (FFT) preprocessing combined with machine learning models works for reliable mechanical component fault detection. Taajobian et al. [6] created a systematic approach for locating faults in an automotive cylinder head by analyzing low-frequency vibrational data. A finite element model (FEM) received simulated controlled impacts then acceleration signals underwent frequency domain transformation to extract fault signatures using statistical reduction methods. The generated signatures helped train a Multi-Layer Perceptron (MLP) that produced fault localization accuracy through reduced sensor requirements and shorter test periods. The research established that fault mapping accuracy depends only on the low-frequency FFT components because it operated under the first natural frequency threshold of 2 kHz. The authors verified this finding in their work.

Van et al. [36] combined FFT with statistical feature extraction to classify gearbox faults. Van et al. [36] converted raw vibration signals to frequency space before applying kurtosis and skewness along with range to compress the FFT signal dimensions. The statistical reduction produced a 32 dimensional feature vector for ANN, LR, and SVM training purposes. The ANN achieved 100% classification accuracy which proved that FFT-derived statistical features maintain high performance in diagnostic applications. The authors demonstrated that statistical pipelines both enhance diagnostic precision and reduce training time, while this research follows the same principle by using low-dimensional high-variance features.

Duc Nguyen et al. [37] established the effectiveness of FFT-driven pipelines by creating an electrical motor diagnostic system that combined frequency domain analysis with shallow machine learning classifiers. The research results supported the idea that FFT-derived features contain vital vibration characteristics which enable fault differentiation regardless of operational variations. Their system achieved high classification rates through signal transformation and ML modeling while maintaining low preprocessing requirements which proved the pipeline's readiness for real-time applications.

Research indicates that FFT signal processing combined with suitable machine learning models produces fault diagnostic solutions which are both accurate and scalable and computationally efficient [6]. The research applies previous findings to develop this work through natural resolution FFT and reduced statistical features and fair linear versus nonlinear model comparison which enhances previous frameworks by adding methodological clarity and model interpretability.

2 Gaps in Literature

2.1 Lack of Simple, Low-Dimensional Feature Sets

Many fault diagnosis studies extract large FFT feature sets often comprising 100+ frequency bins or raw spectral vectors without sufficient dimensionality reduction [30, 31]. Although Taajobian et al. [6] applied statistical filtering to reduce their feature set to 16 components, this still imposes a moderate computational burden and potential overfitting risk when training AI models on limited datasets.

In contrast, this study shows that just five high-variance frequency components, selected using the "SAMPLE" variance formulation, can preserve predictive power while significantly reducing model complexity. This low-dimensional representation makes the approach lightweight, faster to train, and more suitable for embedded or real time diagnostic systems.

2.1.1 Few Studies Compare Linear Versus Nonlinear Models

The overwhelming majority of literature [6] included focuses exclusively on nonlinear models such as MLPs or SVMs. While these methods offer flexibility, they lack transparency and are harder to validate in safety critical applications. The absence of comparative evaluations between linear baselines (e.g., Least Squares Estimation-LSE) and more complex models makes it difficult to quantify the added value of nonlinearity.

This research fills that gap by implementing both LSE and MLP on the same dataset, using identical features and testing sets. The comparison based on Mean

Absolute Error (MAE) clearly demonstrates that while linear models are interpretable, they are less accurate when mapping nonlinear relationships in vibration features.

2.1.2 Overuse of Zero Padding and Artificial Frequency Enhancement

Zero padding is commonly applied to vibration signals before FFT to interpolate between discrete frequency points. While it can improve spectral visualization, it introduces artificial frequency components that do not correspond to the actual dynamics of the system. As noted by Luo et al. [20], zero padding may generate misleading peaks that reduce the physical interpretability of FFT results particularly in fault diagnosis where real spectral signatures are important.

This study avoids such distortions by adhering to the natural frequency resolution determined by the signal length (50 Hz), ensuring that only physically meaningful features are used in model training. This decision improves the credibility of the diagnostic process and avoids over representing noise or non-existent harmonics.

2.1.3 Limited Focus on Real Engine Components

Many studies validating AI + FFT pipelines have focused on laboratory scale systems like beams, rotors, or synthetic datasets [34, 35]. While these are valuable for algorithm benchmarking, they may not capture the material, boundary, or geometric complexities of real-world automotive parts. Taajobian et al. [6] stand out by applying their method to a simulated cylinder head, but the approach remains underexplored in practical diagnostics literature. The current study contributes by analyzing simulated vibrational signals from a full scale cylinder head using realistic fault geometries and FEA generated data. It demonstrates that low-frequency FFT signatures are sufficient for early fault detection in this critical engine component.

3 Summary of Literature Review

The literature review demonstrates that vibration-based fault diagnosis has gained importance through the combination of FFT and machine learning methods. The research has shown that FFT can extract periodic features from structural anomalies and neural networks can perform effective pattern recognition. The current research has several ongoing challenges that need to be addressed.

The majority of feature extraction pipelines depend on extensive high-dimensional input spaces which create training difficulties and increase the risk of overfitting. The literature shows a preference for using complex nonlinear models without testing simpler alternatives which makes it difficult to determine if the added complexity provides any benefits. The practice of using zero padding techniques remains widespread without proper validation because research shows that these methods create artificial distortions which affect the analysis results. The existing literature mainly investigates theoretical mechanical systems and small-scale components which leaves a significant research gap for automotive component analysis.

This research solves these problems through an efficient method which combines physical validation with comparative analysis. The study implements Natural Resolution FFT to prevent artificial frequency interpolation and Statistical Sample Variance to choose five essential frequency features while using LSE and MLP models with identical input data. The research implements this method on a simulated cylinder head model which represents actual structural conditions and serves as an appropriate platform for AI-based diagnostic testing.

The proposed method enhances the field by delivering an affordable and understandable solution for early fault detection which connects theoretical research to practical automotive diagnostic applications.

4 System Description and Methodology

4.1 Overview of the Workflow

This study follows a structured analytical pipeline for detecting faults in a simulated automobile cylinder head using vibration data and machine learning. Although the physical geometry and fault simulation were not conducted first hand by the author, all relevant parameters including excitation force, fault locations, and sensor position were provided by prior FEM-based studies. Based on the supplied simulation data (time versus acceleration), a series of processing steps were implemented in MATLAB to extract meaningful features and train regression models.

The overall workflow, depicted in Fig. 1, consists of the following main stages.

FFT Transformation using natural resolution (no zero padding).

Fault signature creation subtracting the FFT amplitude of the faulty specimen with the healthy specimen (FFT_faulty − FFT_healthy).

Variance Calculation across all faulty cases to select informative frequency bins.

Feature Selection: Retaining only the top 5 frequency bins with the highest variance.

Model Training and Evaluation: Training linear and nonlinear models with the same training set and use the model on the fixed test set.

Evaluate Models using Mean Absolute Error (MAE): Compare Least Squares Estimation (LSE) with Multi-Layer Perceptron (MLP) using MAE as the performance metric.

This workflow allows fault detection to be interpretable and provides a foundation for comparative analysis between linear and nonlinear models.

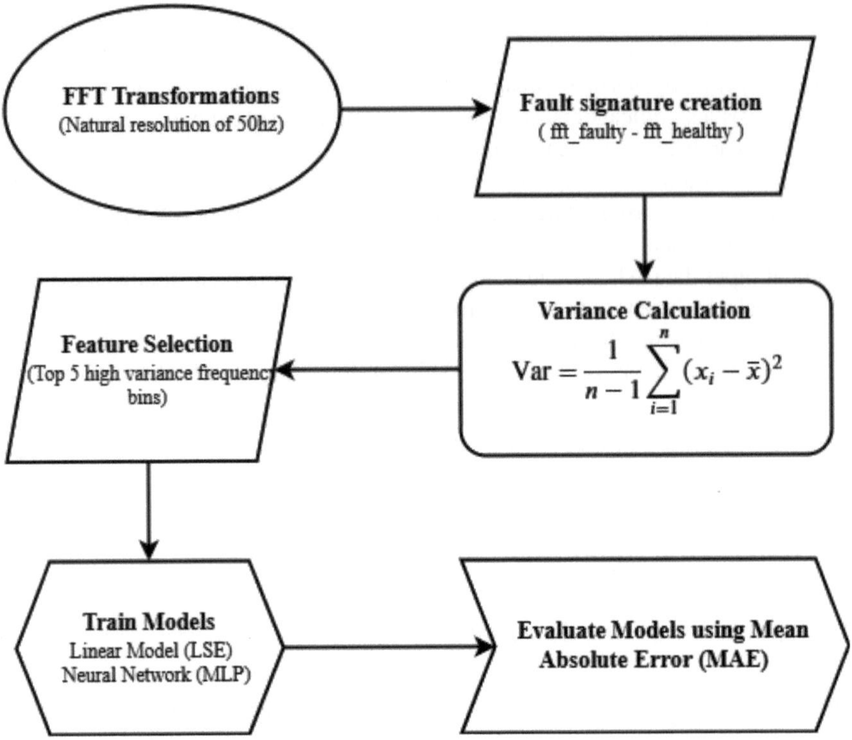

Fig. 1 Summary of workflow

4.2 Vibration Data Origin and Preprocessing

The vibration data used in this study were provided from a previously conducted finite element simulation (FEM) of a cylinder head structure, inspired by the methodology of Taajobian et al. [6]. The simulations involved applying a short duration step force to the structure and measuring acceleration at a fixed sensor point on the surface shown in Fig. 2.

Key parameters of the FEM include:

Excitation force: 100 N.
Sensor sampling rate: 10 kHz.
Signal duration: 0.02 s (200 samples).
Fault locations: 27 distinct positions along a specified axis on the cylinder head shown in Fig. 3.

The dataset comprised 28 cases (1 healthy + 27 faulty), each providing a time series of acceleration versus time. These time domain signals served as the input to the frequency domain processing architecture described in Sect. 2.3.

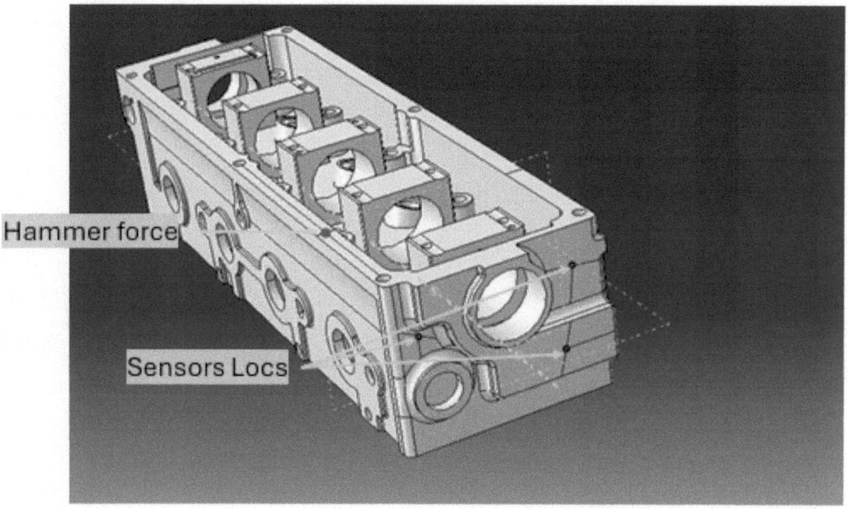

Fig. 2 Hammer force and sensor locations (consistent throughout all tests)

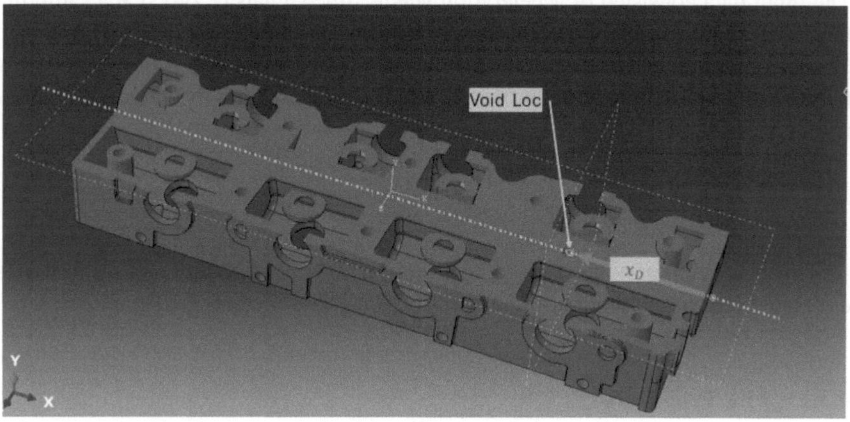

Fig. 3 Fault locations in a single axis, x-axis (single fault per test case, uniform dimension across 27 faulty tests)

4.3 FFT Feature Extraction

Fast Fourier Transform (FFT) was applied to convert the time domain acceleration signals into their frequency domain representations, enabling identification of fault-sensitive vibrational frequencies. The FFT transformation was essential to detect periodic signal variations that occur due to structural faults in the simulated cylinder head.

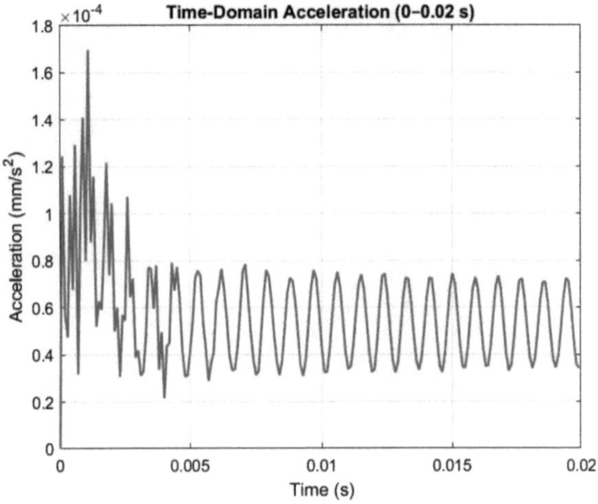

Fig. 4 Time domain acceleration signal (sample case: test P0)

4.3.1 Time Domain Signal Characteristics

The input data consisted of acceleration versus time signals recorded at a sampling rate of 10 kHz for each of the 28 cases (1 healthy, 27 faulty). Each signal contained exactly 200 samples covering a total duration of 0.02 s. This duration was selected to match the simulation output provided, and it directly defines the natural frequency resolution of the FFT:

$$\Delta f = \frac{1}{T} = \frac{1}{0.02\ \mathrm{s}} = 50\ \mathrm{Hz} \tag{1}$$

This resolution provides a physically interpretable representation of the system's dynamic behavior, without artificially introducing intermediate frequencies.

4.3.2 FFT Procedure

Each 200 point acceleration signal was transformed using MATLAB's built in fft() function [38]. The resulting FFT amplitudes were normalized and converted to a single sided spectrum using standard processing techniques. The frequency bins covered the range from 50 to 2000 Hz, producing 40 frequency bins per case. No zero padding was used in the final analysis, which ensures that the frequency components reflect true dynamic behavior.

This method follows best practices in signal processing, as it avoids spectral interpolation that can introduce non-physical artifacts.

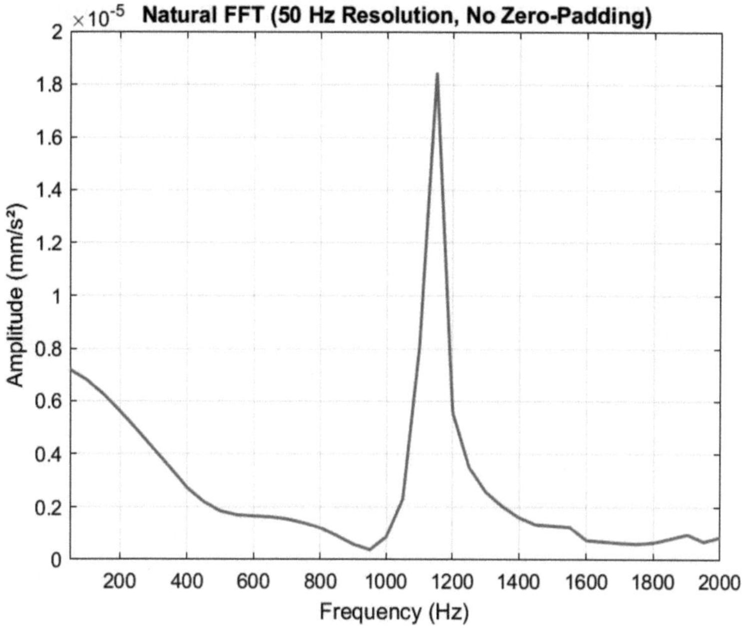

Fig. 5 FFT spectrum (natural resolution) of healthy case (P0)

4.3.3 Comparison with Zero-Padded FFT

To illustrate the effects of zero padding, a secondary FFT was performed with the same signal zero-padded to produce a 10 Hz resolution. While the zero-padded FFT yielded a smoother-looking spectrum, it also introduced fine-scale fluctuations that do not correspond to actual measured dynamics.

The comparison is shown in Fig. 6, where the zero-padded FFT exhibits numerous peaks, while the natural resolution FFT maintains fewer, more physically meaningful components.

This result aligns with the findings of Luo et al. [20], who cautioned that zero padding may create spurious frequency structures that reduce diagnostic clarity. As such, the natural resolution FFT was selected as the primary approach for this study.

4.4 Variance-Based Feature Selection

After FFT transformation, each of the 27 faulty test cases (S1–S27) produced a 40 dimensional frequency spectrum. To identify which frequency bins were most sensitive to fault location changes, statistical variance analysis was applied across all cases.

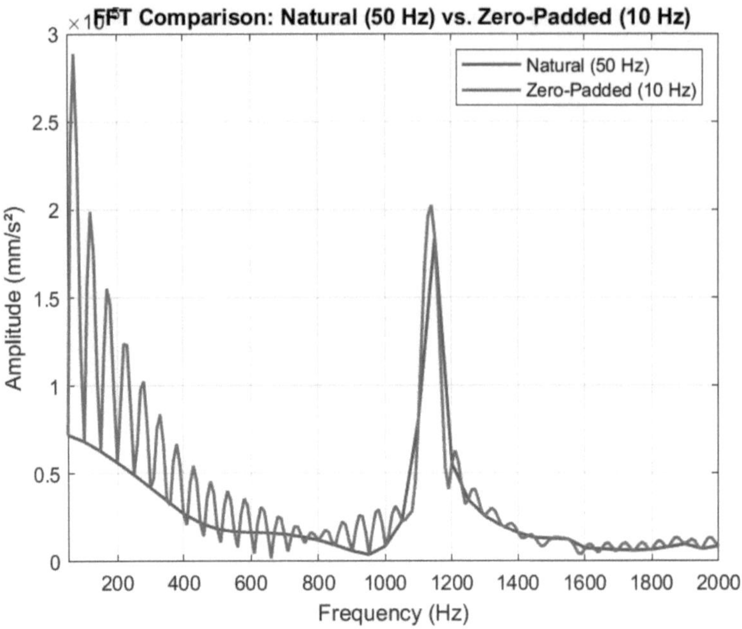

Fig. 6 Comparison of FFT output: zero-padded versus natural resolution

4.4.1 Variance Methodology

For each frequency bin, the sample variance was calculated across all 27 fault cases using the unbiased estimator (Sample Variance, as shown in Fig. 7):

$$\mathrm{Var} = \frac{1}{n-1}\sum_{i=1}^{n}(x_i - \overline{x})^2 \tag{2}$$

where $n = 27$, x_i is the FFT amplitude at a given frequency bin for case i, and $\overline{x}$ is the mean across all 27 cases. The use of $n-1$ corrects for bias in small sample sizes. This approach was also employed by Taajobian et al. [6] for similar feature selection purposes.

4.4.2 Selection of Informative Frequencies

The frequency bins with the five highest variance values were selected as the input features for machine learning. These frequencies were assumed to carry the strongest indicators of presence of faults at different locations. Table 1 lists the selected bins in descending order of variance.

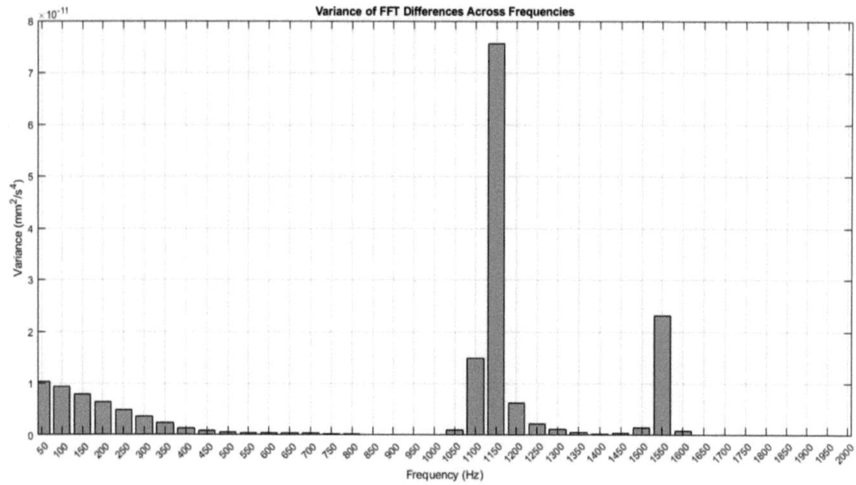

Fig. 7 Variance distribution across frequency bins (50–2000 Hz)

Table 1 Top 5 frequency bins by variance

Rank	Frequency (Hz)	Variance (mm^2/s^4)
1	1150	7.56e−11
2	1550	2.30942e−11
3	1100	1.48712e−11
4	50	1.03734e−11
5	100	9.36037e−12

These five bins formed the five-dimensional input for the machine learning models. Their selection ensured minimal dimensionality while keeping key diagnostic information, helping to reduce overfitting and training time.

4.5 *Machine Learning Model Implementation*

This section describes the implementation of two supervised learning models: a linear regression model using Least Squares Estimation (LSE) and a nonlinear Multi-Layer Perceptron (MLP) neural network. Both models were trained using five high-variance FFT-based frequency features (Sect. 2.4) extracted from vibration signals to predict fault locations in the cylinder head.

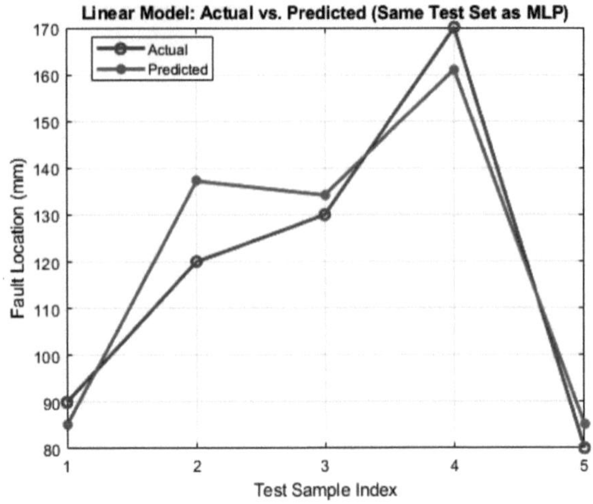

Fig. 8 Linear model–actual versus predicted fault positions

4.5.1 Linear Model (LSE)

The linear model was developed using MATLAB's "newlind()" function, which implements a closed-form Least Squares Estimation (LSE) method based on the formula:

$$\hat{h} = \left(X^{\mathrm{T}}X\right)^{-1}X^{\mathrm{T}}y \tag{3}$$

where X is the matrix of input features and y is the output vector of fault locations. LSE provides an optimal solution for linear regression tasks by minimizing the squared error [39]. Its use is well-supported in both traditional vibration analysis and neural network-based hybrid modeling studies [40].

The feature matrix (27 samples $\times$ 5 features) was normalized per feature and transposed appropriately. A fixed test set consisting of samples (9, 12, 13, 17, and 8) was used to ensure a fair comparison with the MLP model. The remaining 22 samples were used for training (Fig. 8).

4.5.2 Multi-layer Perceptron (MLP)

The MLP model was implemented using MATLAB's "feedforwardnet()" function with a single hidden layer [41]. The architecture included 5 input neurons, 11 hidden neurons, and 1 output neuron. The number of hidden neurons was selected using the heuristic

$$h = 2s + 1 \tag{4}$$

where $s = 5$ is the number of input features [6].

Table 2 MLP hyperparameters

Parameter	Value
Hidden neurons	11
Activation functions	Tansig/purelin
Training algorithm	Levenberg–Marquardt
Initialization	Nguyen–Widrow
Epochs	500
Loss function	MSE
Validation split	Manual (4 samples)

The hidden layer used a tansig activation function, and the output layer used purelin, which is suitable for regression. The model was trained using the Levenberg–Marquardt algorithm, known for its fast convergence in small networks [42]. Weight initialization was performed using the Nguyen–Widrow method. A manual data split was adopted with as mentioned in Table 2:

Training set: 18 samples.
Validation set: 4 samples.
Test set: 5 samples (same as used for LSE).

4.6 Model Evaluation

To quantify model performance, Mean Absolute Error (MAE) was used as the primary evaluation metric. MAE was chosen due to its intuitive interpretation and reduced sensitivity to outliers, making it ideal for real-world diagnostics. It is defined as:

$$\mathrm{MAE} = \frac{1}{n}\sum_{i=1}^{n}\left|y_i - \widehat{y_i}\right| \tag{5}$$

Table 3 summarizes the results obtained from both the LSE and MLP models. The MLP achieved significantly lower MAE, reaffirming the advantage of nonlinear modeling in fault detection tasks. Similar trends were reported by Taajobian et al. [6], where neural networks had high accuracy, and it demonstrates superior generalization capabilities compared to linear approaches in this vibration-based diagnostics.

Table 3 MAE results comparison between LSE and MLP models

Model	MAE (mm)
Linear (LSE)	8.098
MLP (Nonlinear)	1.362

5 Results and Discussion

5.1 Selection and Relevance of Evaluation Metric

For this study, the Mean Absolute Error (MAE) was chosen as the primary evaluation metric due to its simplicity, intuitive interpretability, and robustness against outliers. Unlike metrics such as Mean Squared Error (MSE), which amplify the effect of large deviations, MAE provides a direct measure of average deviation in physical units (millimeters), making it particularly meaningful in real-world fault localization tasks.

5.2 Model Comparison: LSE Versus MLP

The two models received evaluation through the same five test samples (9, 12, 13, 17 and 8). The target variable was the fault position along a 270 mm-long cylinder head.

The LSE model provided MAE of 8.098 mm with 2.999% average error across the entire length. The model offered basic prediction capabilities yet its results showed substantial deviations in particular instances as displayed in Fig. 8. The model fails to capture complex nonlinear patterns in vibration-based FFT features because it operates as a linear system [6, 40].

The MLP model, demosntrated in Fig. 9, achieved MAE of 1.362 mm through its nonlinear activation functions and deep learning capabilities which resulted in a percentage error of 0.504%. The actual and predicted fault positions demonstrate a close match in Fig. 10. The MLP achieves better results because of its mathematical structure shown Fig. 9, effective training with the Levenberg–Marquardt algorithm and manual data splitting.

The MLP model demonstrates stable performance through its smooth training and validation convergence as shown in Fig. 11.

6 Summary of Key Observations

The Mean Absolute Error (MAE) served as a strong evaluation metric because it provides straightforward interpretation of prediction accuracy in millimeters. The Linear Least Squares Estimation (LSE) model provided computational benefits through its closed-form solution yet its performance suffered because it failed to detect the nonlinear patterns in vibration data. The Multi-Layer Perceptron (MLP) model demonstrated better generalization and prediction abilities through its lower absolute errors and decreased error percentages. The combination of Fast Fourier Transform (FFT) frequency domain features with an appropriate learning model such

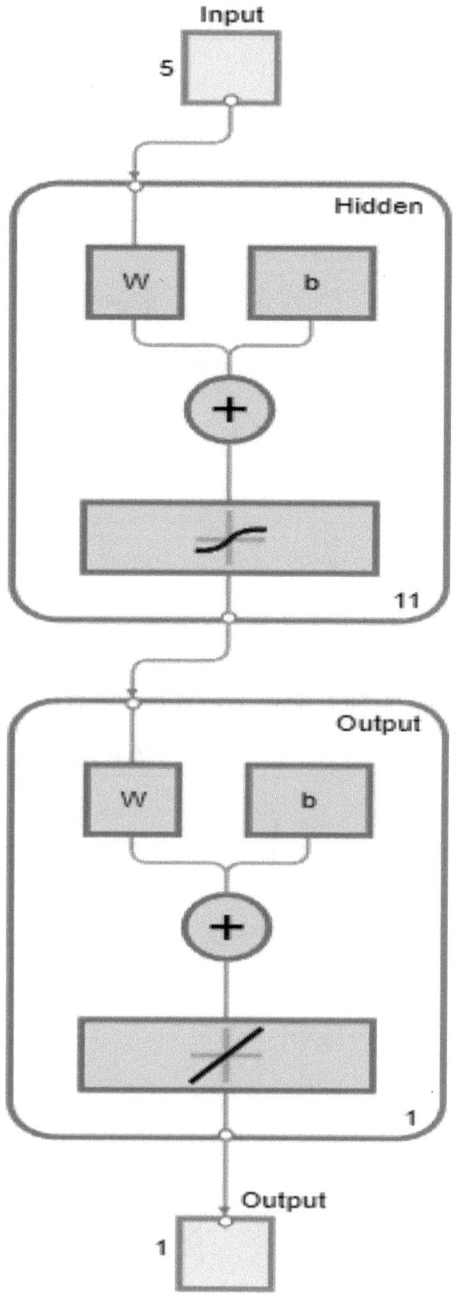

Fig. 9 MLP architecture (5–11–1)

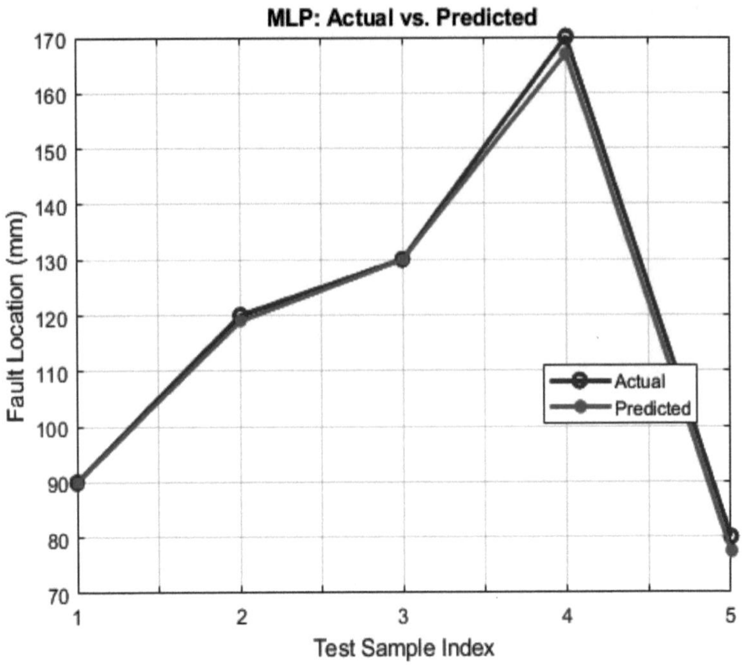

Fig. 10 MLP–actual versus predicted fault positions

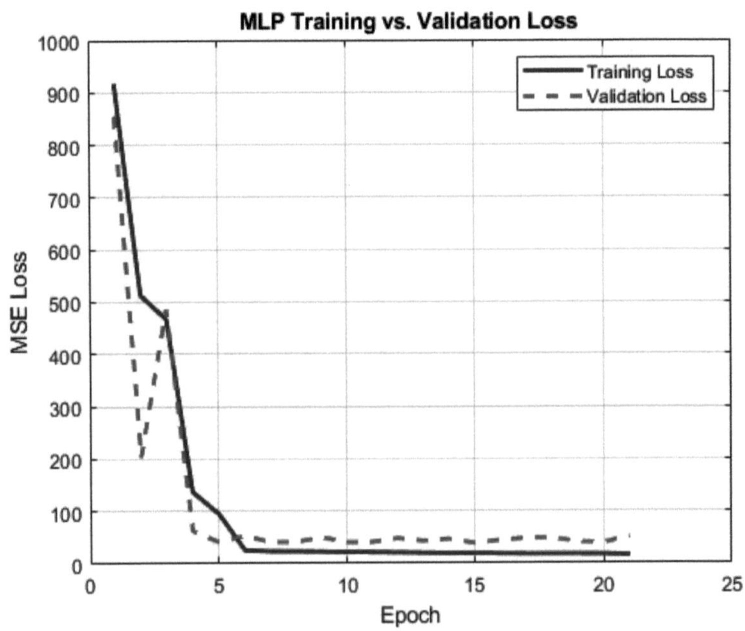

Fig. 11 MLP training versus validation loss

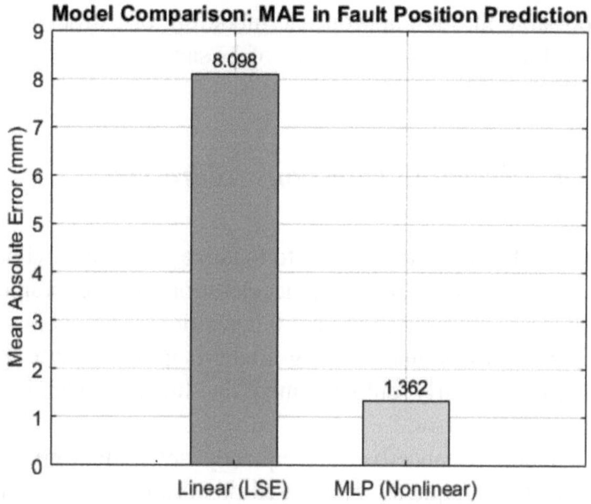

Fig. 12 MAE results comparison–LSE versus MLP

as the MLP resulted in precise fault localization that reached sub-millimeter accuracy levels. The interpretation and communication of model effectiveness became easier through the use of actual versus predicted value plots (Fig.10) and comparative MAE bar charts (Fig. 12).

7 Conclusion

The research employed vibration analysis in frequency domains to detect automotive cylinder head faults through supervised learning techniques. The approach involved extracting valuable information from acceleration signals through Fast Fourier Transform (FFT) and selecting high-variance frequency components before applying both linear model and nonlinear predictive model.

The research introduced variance analysis to select important frequency bands between 50 and 2000 Hz. Two machine learning models were developed for regression-based fault localization: a Least Squares Estimation (LSE) linear model and a Multi-Layer Perceptron (MLP) neural network.

The MLP model achieved superior results than the LSE model by producing a Mean Absolute Error (MAE) of 1.362 mm, while the LSE model produced 8.098 mm. Nonlinear modeling demonstrates superior performance than linear modeling when analyzing vibration-based fault data because it reduces average positional error by five times. The fixed test set evaluation provided a fair and unbiased comparison while training convergence behavior showed good model generalization.

The research demonstrates that frequency domain features extracted from controlled simulation data can function as successful indicators to detect localized structural damage. The research demonstrates that meaningful predictions become possible through appropriate modeling of extracted frequency features even when working with a small dataset of 27 samples.

7.1 Research Questions Addressed

Can vibration data be used to localize faults in a cylinder head?

Yes. FFT-transformed acceleration data captured fault signatures that were successfully used to regress the position of defects.

Which machine learning model is more suitable for this task? The MLP, neural network clearly outperformed the linear model in terms of both absolute and percentage error.

Does variance-based frequency selection improve model performance? Yes. Selecting frequency bins based on variance ensured that only the most informative features were used, which improved prediction accuracy and MLP outperformed the Liner model.

7.2 Future Work

The outcomes of the study are promising though they are limited by the size of available data that the analysis needs to rely on simulated data. In order to make the findings more reliable and applicable, further research may involve the expansion of the dataset with the experimental measurements taken in real life conditions over a variety of operation conditions. Also, there is a possibility to investigate other methods of feature extraction, e.g., wavelet transforms, which can cover more detailed features of the signal. Moreover training different neural network models and comparing their results would allow the use of the best model. Furthermore simulating with different fault modes (crater wear, cracks, etc.) and a combination of them in all 3-axes. Lastly, implementation of the cross-validation strategies would give greater evaluation of the generalizability of the models and minimize the chance of overfitting.

Appendices

Appendices 1–4 were all performed in MATLAB.

Appendix 1: FFT on Acceleration Data

```
%% Load Acceleration Data
% Read 200 acceleration samples from column 5 of 'Sensor1' sheet
[~, ~, raw] = xlsread('P0.xlsx', 'Sensor1');
acc = cell2mat(raw(1:200, 5));   % Extract column 5 (acceleration)

%% FFT Computation with Natural Resolution
Ts = 0.0001;                 % Sampling interval (s)
Fs = 1 / Ts;                 % Sampling frequency (Hz)
N = length(acc);             % Number of samples
df = Fs / N;                 % Frequency resolution (Hz)
f = (0:N/2) * df;            % Frequency vector (one-sided)

Y = fft(acc);                % Compute FFT of the signal
P2 = abs(Y) / N;             % Normalize magnitude spectrum
P1 = P2(1:N/2+1);            % Take one-sided spectrum
P1(2:end-1) = 2 * P1(2:end-1);  % Double non-DC/non-Nyquist
components

%% Select Frequencies up to 2000 Hz
valid_idx = f <= 2000;     % Logical index for f ≤ 2000 Hz
f_out = f(valid_idx);    % Frequencies column vector
amp_out = P1(valid_idx);% Amplitudes column vector

%% Export to Excel File
output_data = [f_out, amp_out];    % Combine frequency and amplitude
xlswrite('FFT_Output_P0.xlsx', output_data, 'Sheet1', 'A1');  %
Write data
```

Appendix 2: Variance Calculation on Fault Signatures

```
% Compute and plot variance across tests for each frequency
component
data = readmatrix('Modified_Differences.xlsx');

accel_values = data(:, 2:28); % Extract S1 to S27 (27 test columns)
frequencies = data(:, 1);    % Extract frequency values (50-2000 Hz)

variance = var(accel_values, 0, 2);  % Row-wise variance across
tests (unbiased estimator)
```

Appendix 3: Least Square Estimation

```
%% Linear Modeling Using Least Squares Estimation (LSE)

% Load feature matrix (5 × 27) and ground-truth fault positions
data = readmatrix('Models_input.xlsx');
X = data(1:5, :);                  % Features: 5 × 27
Y = (1:27)';                       % Fault locations (1 to 27 cm)

% Normalize each feature row-wise, then transpose to 27 × 5
X_norm = normalize(X, 2);
X_norm = X_norm.';                 % Now samples × features: 27 × 5

% Define fixed test and training indices (same as MLP split)
test_set  = [9, 12, 13, 17, 8];
train_set = setdiff(1:27, test_set);

% Split into training and testing sets
X_train = X_norm(train_set, :).';        % 5 × 22
Y_train = Y(train_set).';                % 1 × 22

X_test  = X_norm(test_set, :).';         % 5 × 5
Y_test  = Y(test_set);                   % 5 × 1

% Train linear model using Least Squares Estimation
LM = newlind(X_train, Y_train);

% Predict fault positions for test set
Y_pred = LM(X_test).';

% Convert fault positions to millimeters
Y_test_mm = Y_test * 10;
Y_pred_mm = Y_pred * 10;
```

Appendix 4: Multi-layer-Perceptron

```
%% Multi-Layer Perceptron (MLP) for Fault Position Prediction

% Load 5 selected features for 27 test samples
data = readmatrix('Models_input.xlsx');
X = data.';                    % Transpose to 27 × 5 (samples × features)
Y = (1:27).';                  % Ground-truth fault positions (1 to 27 cm)

% Normalize features column-wise (zero mean, unit variance)
X_norm = normalize(X);

% Defining manual split for training, validation, and testing
train_set = [3, 15, 22, 16, 23, 11, 24, 19, 21, 1, 7, 20, 27, 4, 26,
6, 10, 5];
val_set   = [25, 18, 14, 2];
test_set  = [9, 12, 13, 17, 8];

% Create and configure feedforward neural network
net = feedforwardnet(11, 'trainlm');
net.layers{1}.transferFcn = 'tansig';            % Hidden layer: tan-
sigmoid
net.layers{2}.transferFcn = 'purelin';           % Output layer:
linear
net.performFcn = 'mse';                          % Loss function: MSE

% Apply manual data split
net.divideFcn = 'divideind';
net.divideParam.trainInd = train_set;
net.divideParam.valInd   = val_set;
net.divideParam.testInd  = test_set;

% Use Nguyen-Widrow initialization
net.initFcn              = 'initlay';
net.layers{1}.initFcn = 'initnw';
net.layers{2}.initFcn = 'initnw';
net = init(net);

% Set training parameters
net.trainParam.epochs       = 500;
net.trainParam.showWindow = false;

% Train the MLP
[net, tr] = train(net, X_norm.', Y.'); % Inputs: 5 × 27, Targets: 1
× 27

% Evaluate on test set
Y_test = Y(test_set);
X_test = X_norm(test_set, :).';
Y_pred = net(X_test).';

% Calculate Mean Absolute Error (in mm)
mae = mean(abs(Y_pred - Y_test)) * 10;        % Convert from cm to mm

% Display result
fprintf('MLP - Mean Absolute Error: %.3f mm\n', mae);
```

References

1. Li Y, Liu J, Zhong G, Huang W, Zou R (2021) Analysis of a diesel engine cylinder head failure caused by casting porosity defects. Eng Fail Anal 127:105498. https://doi.org/10.1016/J.ENGFAILANAL.2021.105498
2. Li S, Mao J, Wang S (2008) Cylinder head fem analysis and its improvement. IFIP Int Feder Inform Process 259:761–768. https://doi.org/10.1007/978-0-387-77253-0_4
3. Tama BA, Vania M, Lee S, Lim S (2022) Recent advances in the application of deep learning for fault diagnosis of rotating machinery using vibration signals. Artif Intell Rev 56(5):4667–4709. https://doi.org/10.1007/S10462-022-10293-3
4. Yan YJ, Cheng L, Wu ZY, Yam LH (2007) Development in vibration-based structural damage detection technique. Mech Syst Sig Process 21(5):2198–2211. https://doi.org/10.1016/J.YMSSP.2006.10.002
5. Hakim SJS, Abdul Razak H, Ravanfar SA (2015) Fault diagnosis on beam-like structures from modal parameters using artificial neural networks. Measurement 76:45–61. https://doi.org/10.1016/J.MEASUREMENT.2015.08.021
6. Taajobian M, Mohammadzaheri M, Doustmohammadi M, Amouzadeh A, Emadi M (2018) Fault diagnosis of an automobile cylinder head using low frequency vibrational data. J Mech Sci Technol 32(7):3037–3045. https://doi.org/10.1007/S12206-018-0606-X/METRICS
7. Li-Tian L, Hong-Yun L, Xiang-Long C, Yong-Min F, Yun-Kui X (2011) Diesel misfire fault diagnosis using vibration signal over cylinder head. Lect Notes Electr Eng 100(4):761–768. https://doi.org/10.1007/978-3-642-21762-3_99
8. Liu J, Shi Y, Zhang X, Xu S, Dong L (2011) Fuel injection system fault diagnosis based on cylinder head vibration signal. Proced Eng 16:218–223. https://doi.org/10.1016/J.PROENG.2011.08.1075
9. Wang C, Zhang Y, Zhong Z (2008) Fault diagnosis for diesel valve trains based on time–frequency images. Mech Syst Sig Process 22(8):1981–1993. https://doi.org/10.1016/J.YMSSP.2008.01.016
10. Barelli L, Bidini G, Buratti C, Mariani R (2009) Diagnosis of internal combustion engine through vibration and acoustic pressure non-intrusive measurements. Appl Therm Eng 29(8–9):1707–1713. https://doi.org/10.1016/J.APPLTHERMALENG.2008.07.025
11. Feng J (2021) Failure analysis of rheocast cylinder head covers of hypereutectic Al–Si alloys. J Fail Anal Prevent 21(2):488–493. https://doi.org/10.1007/S11668-020-01110-6/FIGURES/8
12. Xu X, Yu Z (2006) Failure analysis of a diesel engine cylinder head. Eng Fail Anal 13(7):1101–1107. https://doi.org/10.1016/J.ENGFAILANAL.2005.07.016
13. Krstic B, Rasuo B, Trifkovic D, Radisavljevic I, Rajic Z, Dinulovic M (2013) Failure analysis of an aircraft engine cylinder head. Eng Fail Anal 32:1–15. https://doi.org/10.1016/J.ENGFAILANAL.2013.03.004
14. Inês Silva M, Malitckii E, Santos TG, Vilaça P (2023) Review of conventional and advanced non-destructive testing techniques for detection and characterization of small-scale defects. Progr Mater Sci 138:101155. https://doi.org/10.1016/J.PMATSCI.2023.101155
15. Li S, Xin Y, Li X, Wang J, Xu K (2019) A review on the signal processing methods of rotating machinery fault diagnosis. In: Proceedings of 2019 IEEE 8th joint international information technology and artificial intelligence conference, ITAIC 2019, pp 1559–1565. https://doi.org/10.1109/ITAIC.2019.8785572
16. Chu T, Nguyen T, Yoo H, Wang J (2024) A review of vibration analysis and its applications. Heliyon 10(5):e26282. https://doi.org/10.1016/J.HELIYON.2024.E26282
17. Burda EA, Zusman GV, Kudryavtseva IS, Naumenko AP (2022) An overview of vibration analysis techniques for the fault diagnostics of rolling bearings in machinery. Shock Vibr 2022(1):6136231. https://doi.org/10.1155/2022/6136231
18. Xiang C, Ren Z, Shi P, Zhao H (2021) Data-driven fault diagnosis for rolling bearing based on DIT-FFT and XGBoost. Complexity 2021(1):4941966. https://doi.org/10.1155/2021/4941966

19. Jalayer M, Orsenigo C, Vercellis C (2021) Fault detection and diagnosis for rotating machinery: a model based on convolutional LSTM, fast Fourier and continuous wavelet transforms. Comput Ind 125:103378. https://doi.org/10.1016/J.COMPIND.2020.103378
20. Luo J, Xie Z, Xie M (2016) Interpolated DFT algorithms with zero padding for classic windows. Mech Syst Sig Process 70–71:1011–1025. https://doi.org/10.1016/J.YMSSP.2015.09.045
21. Yang Y, Liu C, Jiang D, Behdinan K (2020) Nonlinear vibration signatures for localized fault of rolling element bearing in rotor-bearing-casing system. Int J Mech Sci 173:105449. https://doi.org/10.1016/J.IJMECSCI.2020.105449
22. Jayaswal P, Wadhwani AK, Mulchandani KB (2008) Machine fault signature analysis. Int J Rotat Mach 2008(1):583982. https://doi.org/10.1155/2008/583982
23. Ayman A, Onsy A, Attallah O, Brooks H, Morsi I (2025) Feature learning for bearing prognostics: a comprehensive review of machine/deep learning methods, challenges, and opportunities. Measurement 245:116589. https://doi.org/10.1016/J.MEASUREMENT.2024.116589
24. Fuchs P, Kröger T, Garbe CS (2021) Defect detection in CT scans of cast aluminum parts: a machine vision perspective. Neurocomputing 453:85–96. https://doi.org/10.1016/J.NEUCOM.2021.04.094
25. Sowmya S, Saimurugan M, Edinbarough I (2024) Rotational machine fault diagnosis using artificial intelligence (AI) strategies for the operational challenges under variable speed condition: a review. IEEE Access. https://doi.org/10.1109/ACCESS.2024.3469212
26. Plante T, Nejadpak A, Yang CX (2015) Faults detection and failures prediction using vibration analysis. Autotestcon 2015:227–231. https://doi.org/10.1109/AUTEST.2015.7356493
27. Mohd Ghazali MH, Rahiman W (2021) Vibration analysis for machine monitoring and diagnosis: a systematic review. Shock Vibr 2021(1):9469318. https://doi.org/10.1155/2021/9469318
28. Betta G, Liguori C, Paolillo A, Pietrosanto A (2002) A DSP-based FFT-analyzer for the fault diagnosis of rotating machine based on vibration analysis. IEEE Trans Instrum Measur 51(6):1316–1321. https://doi.org/10.1109/TIM.2002.807987
29. Al-Badour F, Sunar M, Cheded L (2011) Vibration analysis of rotating machinery using time–frequency analysis and wavelet techniques. Mech Syst Sig Process 25(6):2083–2101. https://doi.org/10.1016/J.YMSSP.2011.01.017
30. Ghaderi H, Kabiri P (2012) Fourier transform and correlation-based feature selection for fault detection of automobile engines. In: AISP 2012—16th CSI international symposium on artificial intelligence and signal processing, pp 514–519. https://doi.org/10.1109/AISP.2012.6313801
31. De Assis Boldt F, Rauber TW, Varejão FM, Ribeiro MP (2015) Fast feature selection using hybrid ranking and wrapper approach for automatic fault diagnosis of motorpumps based on vibration signals. In: Proceeding of the 2015 IEEE international conference on industrial informatics, INDIN 2015, pp 127–132. https://doi.org/10.1109/INDIN.2015.7281722
32. Agustin XM, Samson EJ, Ostia C (2023) Application of SVM classification technique in single-phase AC motor bearing fault diagnosis using motor current analysis with MRA: FFT as feature extractor and CFFS as feature selection method. In: Proceedings of the 2023 IEEE 14th control and system graduate research colloquium, ICSGRC 2023—conference proceeding, pp 208–213. https://doi.org/10.1109/ICSGRC57744.2023.10215397
33. Ghate VN, Dudul SV (2010) Optimal MLP neural network classifier for fault detection of three phase induction motor. Exp Syst Appl 37(4):3468–3481. https://doi.org/10.1016/J.ESWA.2009.10.041
34. Goyal D, Dhami SS, Pabla BS (2020) Non-contact fault diagnosis of bearings in machine learning environment. IEEE Sens J 20(9):4816–4823. https://doi.org/10.1109/JSEN.2020.2964633
35. Liyanagedera ND, Ratnaweera A, Randeniya DIB (2013) Vibration signal analysis for fault detection of combustion engine using neural network. In: Proceedings of the 2013 IEEE 8th international conference on industrial and information systems, ICIIS 2013—conference proceedings, pp 427–432. https://doi.org/10.1109/ICIINFS.2013.6732022

36. Van B, Van Hoa N, Nguyen H, Jang YM (2020) Statistical feature extraction in machine fault detection using vibration signal. In: International conference on ICT convergence, 2020-October, pp 666–669. https://doi.org/10.1109/ICTC49870.2020.9289285
37. Duc Nguyen V, Zwanenburg E, Limmer S, Luijben W, Back T, Olhofer M (2021) A combination of Fourier transform and machine learning for fault detection and diagnosis of induction motors. In: Proceedings of the 2021 8th international conference on dependable systems and their applications, DSA 2021, pp 344–351. https://doi.org/10.1109/DSA52907.2021.00053
38. FFT (n.d.) Fast Fourier transform—MATLAB. https://www.mathworks.com/help/matlab/ref/fft.html. Accessed 18 June 2025
39. Newlind (n.d.) Design linear layer—MATLAB. https://www.mathworks.com/help/deeplearning/ref/newlind.html. Accessed 18 June 2025
40. Mohammadzaheri M, Chen L, Ghaffari A, Willison J (2009) A combination of linear and nonlinear activation functions in neural networks for modeling a de-superheater. Simul Model Pract Theory 17(2):398–407. https://doi.org/10.1016/J.SIMPAT.2008.09.015
41. Feedforwardnet (n.d.) Generate feedforward neural network—MATLAB. https://uk.mathworks.com/help/deeplearning/ref/feedforwardnet.html. Accessed 18 June 2025
42. Demuth H, Beale M (2001) Neural network toolbox for use with MATLAB: user's guide, version 4

Development of a Data-Driven Predictive Maintenance Framework for Centrifugal Pumps

Lubayd Bahwan, Hamed Al Harthy, Morteza Mohammadzaheri, and Johaness Earlings

Abstract Centrifugal pumps play critical roles in industries such as oil and gas, water treatment, and power generation. Traditional pump maintenance approaches, whether reactive or preventive, are often inefficient, which results in unplanned downtime and increased operational costs. This chapter examines the application of artificial intelligence (AI) for predictive maintenance (PdM) of centrifugal pumps. It uses real-world vibration data to estimate the remaining useful life (RUL) and predict failures before they happen. A comprehensive literature review highlights the limitations of traditional methods, including acoustic emission, motor current signature analysis, and vibration analysis. It also presents the current trends in machine learning, edge computing, and explainable AI that allow for more informative and meaningful performance-based maintenance decisions. The methodology we used includes acquiring and preprocessing the time-domain vibration signals taken from six pumps of different health conditions. A new metric, DaysToFailure, is formulated to function as a surrogate target in the absence of real failure timestamps. For feature construction, a seven-day sliding window is used with average and range values of time-domain vibration values. A feedforward neural network was trained using Bayesian regularization to prevent overfitting. In our analysis, the mean squared error was reduced from 1400 to about 300, indicating a significant performance improvement. The model can provide a high predictive accuracy on healthy pumps and reasonable generalization in noisy and faulty conditions. Remarkably, the model can identify the impacts of the maintenance events based on the differences in the vibration occurring data, despite the lack of their explicit labels or timestamps, thus demonstrating itself to be suitable for practical industrial use. The results validate the notion that pump life can be elongated, unplanned failures can be avoided, and smart maintenance planning can be improved using an AI-based predictive maintenance structure. This research contributes to the development of scalable, interpretable PdM systems. These systems can be integrated into existing industrial monitoring infrastructures. Future studies are proposed for the fields of frequency-domain analysis, SCADA integration into PdM, and multi-sensor data fusion for PdM. With these new

---

L. Bahwan · H. Al Harthy · M. Mohammadzaheri (✉) · J. Earlings
German University of Technology in Oman, Athaibah, Sultanate of Oman
e-mail: morteza.zaheri@gutech.edu.om

A. Azizi (ed.), *Smart Diagnostics and Predictive Maintenance*, Emerging Trends in Mechatronics, https://doi.org/10.1007/978-981-95-3921-5_5

features, fault detection will be faster, less downtime will occur, and pump maintenance can be better organized based on more dependable and relevant information. As a result, companies will see better performance from their processes and cut down on the cost of repairs.

Keywords Artificial intelligence · Centrifugal pumps · Predictive maintenance · Artificial neural networks · Bayesian regualtion

1 Introduction and Literature Review

1.1 *Background and Practical Significance*

Industrial processes such as water or wastewater treatment, power generation, oil and gas production, and chemical processing are impossible without centrifugal pumps. For fluid transport to flow smoothly, it is important that the pumps function correctly and safety hazards and costs are kept low. While pump failures may be unexpected, it is more effective to conduct preventive maintenance as opposed to reactive maintenance which tends to be expensive. Given that centrifugal pumps play a crucial role, their breakdown can put the operations probability at risk, not just the engineering process.

Centrifugal pumps have been developed over the course of centuries from rudimentary mechanical devices to highly sophisticated machines which are used in modern industrial operations. With advancements in engineering and material science, pumps have gradually increased in performance, strength, and efficiency. Although these advances have reached the present, the risk of an unexpected breakdown is still not prevented, creating a dilemma that the latest industries cannot afford. The above discussion inspires the research of recent maintenance techniques designed to elongate the life of the equipment.

1.2 *Predictive Maintenance by Means of Artificial Intelligence*

For a long time, sensors were used for condition monitoring, but first-time solutions were still limited, and people interpreted the data manually. Thanks to advancements in sensor technology, system performance monitoring data can now be more accurate and abundant. In addition, artificial intelligence has also made it possible for interpretation to be done faster and more accurately with a capacity to store the data otherwise not available to humans. The improvements achieved from these developments enable the possibilities of AI predictive maintenance that not only detects anomalous condition in real time but also keeps a historical database of it, which would continuously learn and improve the predictive capabilities, and give a real advantage over traditional monitoring.

1.3 Chapter Objectives and Scope

The chapter looks at applying predictive maintenance methods to centrifugal pumps used in industry instead of focusing on how pump failures might be evaluated by the systems performance. In addition, it examines the use of AI-driven predictive maintenance systems as an innovative strategy to resolve some of these problems. The study is focused on linking past ways of maintenance work with new predictive solutions. Determining what starts centrifugal pump failures was the point around which this study focused. This involved choosing the AI model that worked best for predictive maintenance in pumps and assessing what value AI-based predictive maintenance offered.

1.4 Chapter Structure

To meet these research objectives, the chapter is divided into various interrelated sections.

- **Literature Review**: This section will give an extensive survey of past research on centrifugal pumps, operations challenges, and currently available maintenance practices. It also discusses recent AI technology progress made in relation to predictive maintenance.
- **Basic System Description and Methodology**: Here, the focus is given in elaborating on the technical details of the given centrifugal pump system under consideration, the design of the related AI algorithms, and the related analysis approach.
- **Results and Discussion**: The outcomes are shared as they show that using AI-based maintenance leads to better and more reliable results than today's strategies. Besides that, this section presents ideas for putting the results into practice, discusses the limitations, and provides suggestions for additional studies.

1.5 Chapter Contribution

Overall, this chapter presents the development of an AI-powered approach for predictive maintenance of centrifugal pumps in the industry. By analyzing published literature, building models using data, and reviewing recorded vibrations, this work advances efficient maintenance, limits unplanned production loss, and enhances how to operate centrifugal pumps.

1.6 Introduction

This chapter aims to examine the progress made in the development of maintenance methods for centrifugal pumps and points out how AI is playing a bigger role in predictive maintenance. It starts with considering the old methods of fault detection, which include acoustic emission and motor current signature analysis, and ends with innovative AI methods which include machine learning, vibration analysis, and using sensors, such as accelerometers and acoustic emission sensors. The review points out current issues and introduces recent technologies that can make the process more dependable and leads to less system failures.

1.7 Cavitation Detection Techniques

In Alfayez and Mba [1], conducted a study to investigate the application of acoustic emission (AE) to detect incipient cavitation and calculate the break-even point (BEP) on centrifugal pumps. This study's importance lies in overcoming the shortcomings of the conventional techniques (including vibration analysis and hydrophone measurement), which often cannot detect the early symptoms of cavitation. The ability of AE to detect high-frequency stress waves makes it a potential replacement for real-time pump condition monitoring by using time–frequency analysis.

The study targeted three centrifugal pumps with different power outputs 60, 127, and 2.2 MW, which were evaluated for performance and net positive suction head (NPSH). AE sensors were positioned on the pump casing as well as the suction and discharge pipes. It was found that AE activity is very closely related to the start of cavitation. This supports the use of AE as a technique for predicting cavitation before an obvious loss of head is established. The findings showed that the AE activity was minimal under the normal BEP conditions and indicating AE's potential to improve pump efficiency. In addition, it was found that the main AE activity takes place at high frequencies.

The study includes many advantages but also has several limitations. Even though cavitation is believed to be the main cause of AE, older research by Neill et al. [2] proved that AE did not always detect signs at the beginning. These examples are referenced in the paper by Alfayez and Mba. The research has not addressed the fact that operating conditions, including the fluids and noise around the structure, can alter the information gathered by AE. Because environmental noises were not considered, there could be some errors in the findings. Further study is necessary to know how AE works and how well it performs in various pumps and under various conditions. The study clearly indicates that AE supports consistent monitoring of pumps and increases their efficiency. It is established that AE is a powerful approach in existing monitoring practices but needs adequate validation and development before it becomes fully compatible with industrial settings.

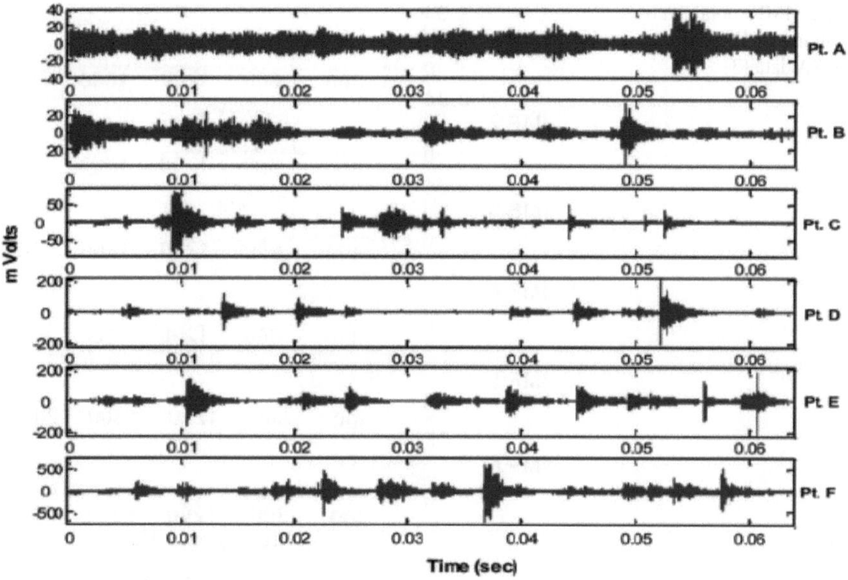

Fig. 1 Frequency spectrum of acoustic emission signals [1]

As shown in Fig. 1, AE data displays changes in emission signals, measured in millivolts (mV) over a period. Pt A represents the standard healthy pump, and Pt B to Pt F show how different pumps operate at various times. There is a gradual increase in the signal strength between Pump A and Pump F, but Pumps D–F show the most obvious bursts, suggestive of more mechanical movement. The signals may pulse at points, but no single dominant frequency is found throughout the data. The plot makes it possible to observe how readings for AE vary and compare between all units.

Sun et al. [3] developed a new method of diagnosing cavitation of centrifugal pumps based on motor current signal analysis (MCSA). It tracks cavitation by observing changes in the current used by the motor. This technique is more effective than previous ones because conventional sensors like pressure pulsation devices are easily affected by noise and are not easy to adjust. The study highlights that MCSA is a workable, non-intrusive real-time tool for detection and localization of cavitation.

To make the results more accurate, the researchers applied complete ensemble empirical mode decomposition applying adaptive noise (CEEMDAN), an algorithm that helps separate different modes in the motor current. The approach processes motor current signals into intrinsic mode functions (IMFs). They provide clues about the features that cause cavitation to form and grow. Different frequencies are detected with each IMF, and some of them overlap with cavitation-caused sounds. The ability to identify time-varying frequencies is drastically improved by using CEEMDAN.

Oscillations occurring from 3 and above on the IMFs indicate that cavitation began, displayed in Fig. 2. The changes in the peak amplitude were followed to

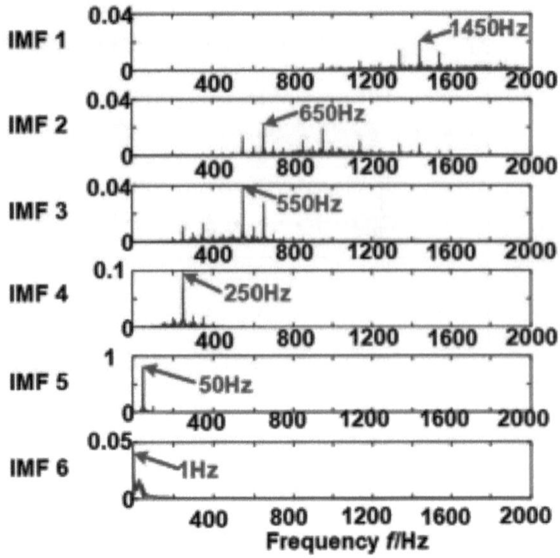

Fig. 2 IMFs obtained by applying CEEMDAN to MCSA data [3]

scan its progress. At the start, early-stage IMFs involve a lot of fast, high-frequency signals, but later, their signals consist mostly of low-frequency parts.

Field studies, conducted using a centrifugal pump, confirm the MCSA technique to be superior to other methods by both measures of accuracy and sensitivity. However, the research does point to several meaningful deficiencies. The performance of MCSA relies on how the pump design addresses change in torque and the behavior of motor currents. Because motor current does not reveal the type of fault, it is difficult to tell cavitation apart from other motor problems. Because this method does not manage the variations of speed well in modern pumps, it cannot be relied on in industries.

Although it focuses on a particular case, the paper demonstrates how MCSA can be used to check on pump operation and to warn of cavitation. With CEEMDAN, you derive results that are both reliable and usable for immediate diagnosis. Currently, MCSA is not suitable for main use in cavitation detection and needs to be updated and tested in industry.

1.8 Centrifugal Pump Failure Modes

McKree et al. [4] conducted a detailed analysis of failure modes in centrifugal pumps with a special emphasis on their application in water supply and sewage systems. Hydraulic, mechanical, and other failures are outlined, and the study reviews their reasons, methods of detection, and how to prevent them. As stated in the paper, condition monitoring helps to cut costs and raise effectiveness at workplaces.

The research studies various hydraulic issues including cavitation, surges in pressure, and flow recirculation. Cavitation due to insufficient net positive suction head (NPSH) is a known problem which causes erosion, noise, and vibration. To reverse the trend, the authors suggest replacing the impeller and restructuring the suction system. Bearing wear, seal failure, and excessive vibration are frequently associated with several factors: misalignment, a lack of lubrication, and poor operating conditions. Both vibration measurement and thermal indicators are emphasized as practical ways of identifying these challenges.

Although sensors have improved, several users continue to overlook the usefulness of machine condition data. According to the authors, predictive maintenance is vital and dependable expert advice and real-time diagnostics can help. The analysis is detailed, but its lack of figures means it is not especially useful for maintenance teams following reliability strategies.

Key reasons for centrifugal pump failures are outlined in the paper, and it urges the use of improved monitoring techniques. Both pump lifespan and reliability are predicted to grow as future tech takes advantage of machine learning and enables pumps to self-check and maintain themselves. According to the study, fast Fourier transform (FFT) should be used to investigate the amplitudes of the signal in the cavitation frequency band.

1.9 Shaft Fatigue and Mechanical Failure

Hua et al. [5] investigated the causes of fatigue failure of a 304 stainless steel shaft in multistage centrifugal pumps. Researchers found that stress buildup and regular bending cause machines to break down in this way. Based on fracture types, scanning electron microscope (SEM) imaging and finite element analysis (FEA) simulations were used to find that substantial amounts of stress cause cracking and inclusions and cavities raise the risk of failure. The researchers studied how to increase the tolerance of stress while using the same materials.

Insufficient fillet radius, a crooked shaft, and out-of-balance impellers are major causes of premature fatigue. To address this, the study suggests the use of 2 mm fillet radius in combination with an elastic coupling. The FEA simulations demonstrated that stress concentrations were substantially reduced by making this modification, as can be seen in Fig. 3. Research findings showed that the improved fillet led to 29% greater fatigue resistance and that an elastic coupling raised the lifespan by over double.

The modern designs with larger fillet diameters enhanced the life of the shaft, but the study has not explored what would happen in factory settings or rather different manufacturing processes. When variable load changes and real-time monitoring are used, the reliability of predictive maintenance methods is increased. Thus, the paper explains what leads to damage on pump shafts and recommends ways to amend the design to resist fatigue which helps ensure both reliability and ease of upkeep.

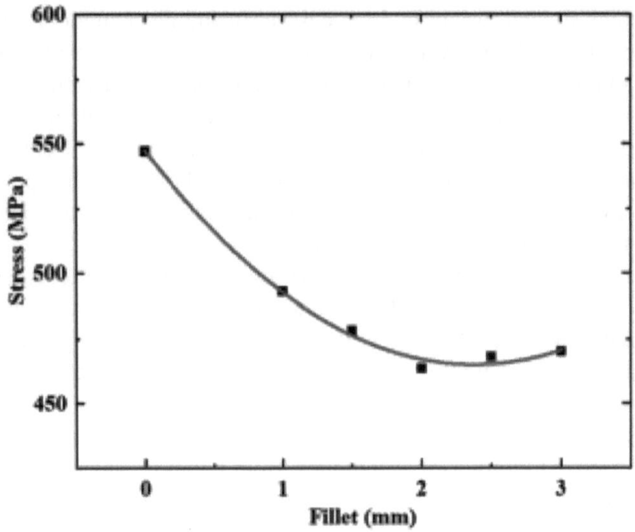

Fig. 3 Stress points at different fillet diameters [5]

1.10 Fault Monitoring Using Motor Current and Vibration

An extensive dataset for fault monitoring of e-motor driven centrifugal pumps using motor current and vibration analysis is presented by Bruinsma et al. [6]. The study contributes value through accurate labeling of data that helps find mechanical and electrical problems and develop ways to prevent them in the future.

The data used included the responses of two centrifugal pumps when they underwent 20 test conditions, including tests with bearing faults, impeller damage, misalignment, cavitation and stator winding shorts, or broken rotor bars. High-frequency accelerometers and current clamps at 20 kHz sampling rate were used to collect the data under different speeds and fault severities. This dataset is shown to be especially useful for training AI models for fault detection and condition-based monitoring. Figure 4 shows the wide range of faults with different operating conditions, which will be useful for training AI models.

A standout strength is the structured and repeatable way of fault simulation that makes it dependable for comparative studies. The data for baseline healthy motors is absent in the cases of bent shafts and broken rotor bars. Even when machines are healthy, there may be some noise introduced in their results from inevitable cavitation. Moreover, vibration and current measurements were not synchronized, which may affect the accuracy of fault diagnosis. The report does not explain the lack of synchronization between the two measurements.

The research paper is a great contribution to the area of predictive maintenance research as it offers a wealth of data for developing AI-driven fault detection models.

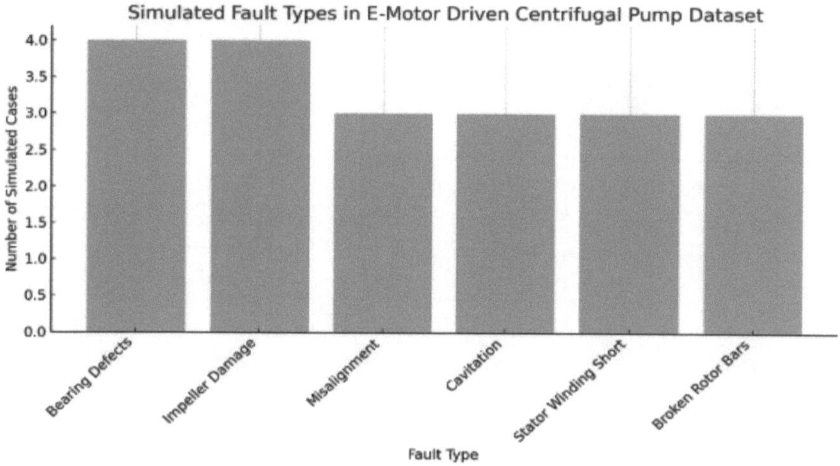

Fig. 4 Simulation setting for vibration and current faults

More work in the future should center on including diverse real-world situations and better data synchronization to prepare the system for factories.

1.11 Data-Driven Pump Failure Prediction

Mohammed [7] presents a multiple linear regression (MLR) model to estimate failure rates of seawater pumps used in the oil and gas settings. The goal of this research is to enhance maintenance strategies through a simple approach for failure prediction using actual operational data. Unlike prior works that needed experimental results, this study employs actual operational data to improve the reliability of failure predictions.

The study recognizes six major factors that cause pump failures as follows: temperature, pressure, mean time between failure, pH level, NH_3 concentration, and total phosphorus in seawater. The proposed model was developed and validated using operational data from the years 2018–2021, which resulted in a 97% success rate for predicting pump failures. The results of sensitivity analysis indicated that failure in the pump was most strongly dependent on pressure and total phosphorus. A graph of the results is present in Fig. 5. Mechanical loads, changes in fluid flow, and an increase in cavitation risk led to problems with pump failure because of pressure.

This study is especially valuable because the data came from practical situations, making it useful to implement. When there are nonlinear links among variables, the MLR is not highly effective, implying that machine learning could find improved answers. In addition, the research emphasizes the need for monitoring pressure surges but does not address issues concerning real-time operations or interactions with predictive maintenance systems.

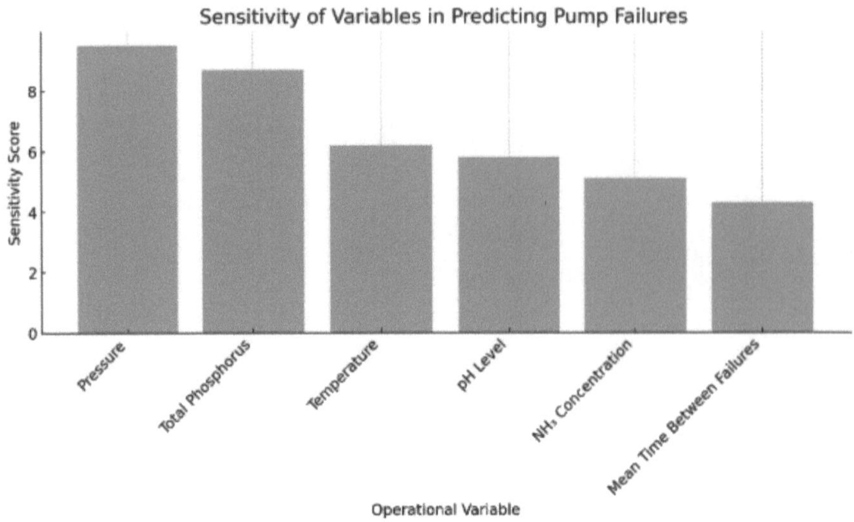

Fig. 5 MLR-based predictions versus actual trends

In summary, the research shows how the use of data can help spot pump breakdowns early and keep equipment well maintained. Future research can contribute to the improvement of the reliability of the models by integrating the methods of machine learning and deploying the systems of real-time monitoring that would increase the predictive accuracy.

1.12 Explainable AI in Predictive Maintenance

Rajaoarisoa et al. [8] introduced a decision-making model that applies Explainable Artificial Intelligence (XAI) for wind turbine predictive maintenance. The work concentrates on using a transparent tool to spot anomalies and analyze when maintenance should take place using machine learning.

A multistage neural network in the model is used to identify anomalies and describe their reasons. Auto-encoder (AE-net), discrete event-based system (DEVS), and recursive neural network (RNN) are used in the methodology for feature extraction, rule-based detection of anomalies, and RUL prediction, respectively. Generator power, angle of the turbine blades, and voltage circuit temperature are key factors studied to maintain turbine health.

With a wind farm dataset from Energias De Portugal, a system based on AE-net-DEVS had 80% accuracy in finding anomalies and 96% accuracy in remaining useful life (RUL) prediction, beating long short-term memory (LSTM) and decision trees. The generator power, blade pitch angle, and voltage circuit temperature were automatically found and extracted by the auto-encoder with unsupervised learning.

Figure 6 details all the steps involved in running a predictive maintenance system: collecting data, choosing notable features, grouping failures into types, calculating RUL, and finally making maintenance choices.

This study stands out due to its clear explainability, offering maintenance operators practical and easily understood advice. A quick shift in generator voltage or increase in the angle of the wind turbine's blades is obvious in the data. However, the assumed system behavior may not be fully represented by a limited set of variables under dynamic industrial environments. Moreover, the choice of the best time window for anomaly detection is not automated, which may influence the model's performance. The cause may be that processing speed, ability to notice noise, and a static sliding time window are all affected by this limitation. Figure 7 shows the accuracy of different models used.

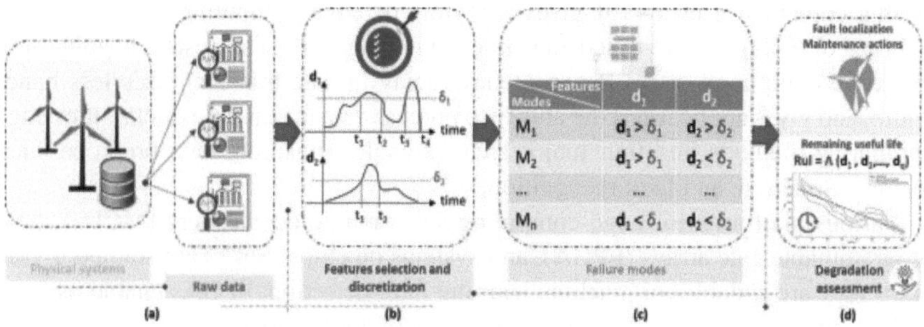

Fig. 6 Auto-encoder structure for wind turbine PdM [8]

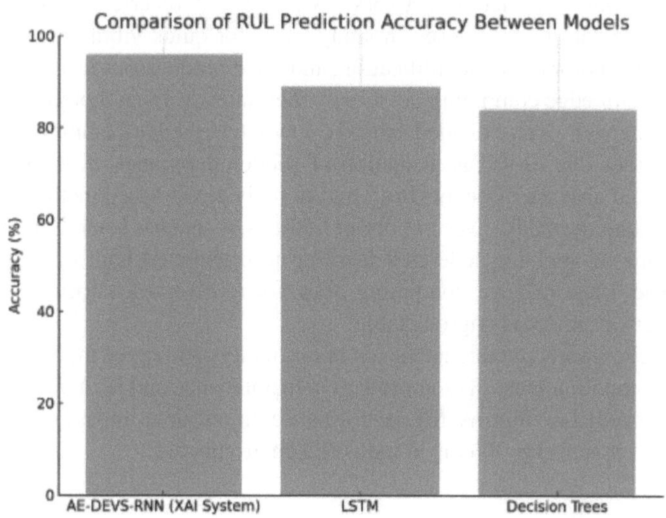

Fig. 7 Accuracy of RUL prediction between models

In general, this research provides an important contribution to predictive maintenance by combining XAI for better decision-making. For further improvements, future research can study instant adjustment models and automatic selection of system features to make the system reliable and easy to use. Researchers have found that generator power, blade pitch angle, and the temperature of the voltage circuit indicate how healthy the turbine is. In real-time, operators monitor these features to check the condition of the mechanical, electrical, and control systems connected to the turbine.

1.13 Edge Computing and IoT in Fault Diagnosis

In this context, Lu et al. [9] gives a review of edge computing applications in Internet of Things (IoT)-based machine fault diagnosis. The research points out that edge computing offers lower latency (delay in data transfer), uses less bandwidth, and gives applications the ability to process data in real time. It classifies edge computing methods into four major areas: signal acquisition, data preprocessing, feature extraction, and fault recognition.

Various platforms for edge computing such as microcontrollers (MCUs), field programmable gate arrays (FPGA), and systems on chips (SoCs) are studied to find out if they are suitable for machine learning fault detection. FPGAs enable users to make their own hardware designs for a variety of applications.

The second part of the study reviews some of the recent developments in lightweight machine learning (ML) and deep learning (DL) models for edge devices such as convolutional neural networks (CNNs) and recurrent neural networks (RNNs). Edge computing is used in study cases for quick vibration signal reduction, real-time motor failure identification, and rotor condition review. Figure 8 ranks distinct types of edge computing devices by their latency. Even though pump degradation can happen over extended periods, low-latency edge computing is vital in these situations due to its fast detection of sudden anomalies, available real-time responses, and absence of delays from traditionally cloud-based methods.

Even though the study gives important findings, it also has some limitations. The practical side of applying federated learning to distributed fault diagnosis is still limited. The usage of edge computing does not receive much focus in situations where the environment is unpredictable.

The article reports notable advances in our understanding of the increasing role that edge computing must play in predictive maintenance and fault diagnosis. More studies are needed to improve ML performance on resource-limited equipment and to apply it to practical problems in industrial environments.

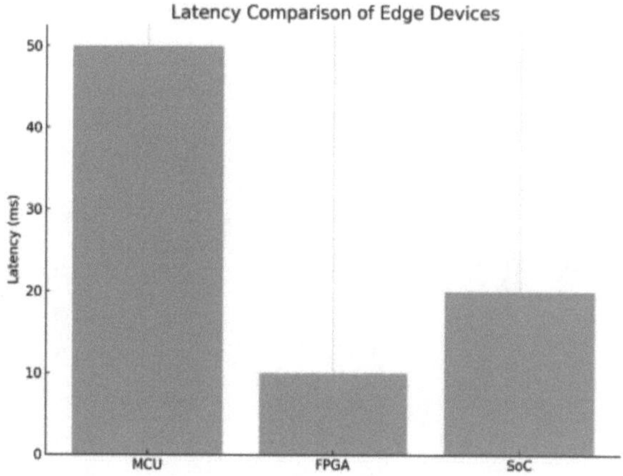

Fig. 8 Comparison of latency between edge devices

1.14 Non-destructive Fault Detection Techniques

In this chapter, Irfan et al. [10] present a new fault detection framework for bearing failure diagnosis of water pumps based on instantaneous power analysis (IPA) and a strong feature selection and extraction algorithm (SFSEA). This study presents a noninvasive analysis technique to detect three separate fault-related harmonics to overcome limitations of the conventional vibration analysis (VA) and motor current analysis (MCA). Three harmonics that appear are the fundamental frequency and fault-related harmonics from the outer and inner race of the bearing.

SFSEA methodology extracts fault-related features from IPA spectrum and feeds these features to an extreme gradient boosting (XGB) classifier. The simulated minor bearing faults—holes and scratches—are introduced on a two-horsepower induction motor. These results are representative for a pump. The results indicate that the machine can identify all nine machine health states correctly, as seen in Fig. 9. It exceeded the accuracy of support vector machines (SVM), k-nearest neighbors (k-NN), and convolutional neural networks (CNNs) which all hit less than 90% of accuracy.

An important advantage of the study is its process of extracting features which remains reliable in industrial settings despite environmental disturbance. But the fact that the research relies on a controlled laboratory means it cannot always be used to solve problems in the real world. The functioning of the IPA signal may differ under different loading and external conditions. In addition, while the SFSEA feature selection algorithm helps boost performance, it has not yet been tried out on a large scale in real-world use.

The research provides introduction to a promising process that can find faults in water pump bearings without invasive testing. Future researchers should confirm the effectiveness of SFSEA across many industrial environments and consider how it could relate to edge computing for rapid diagnoses.

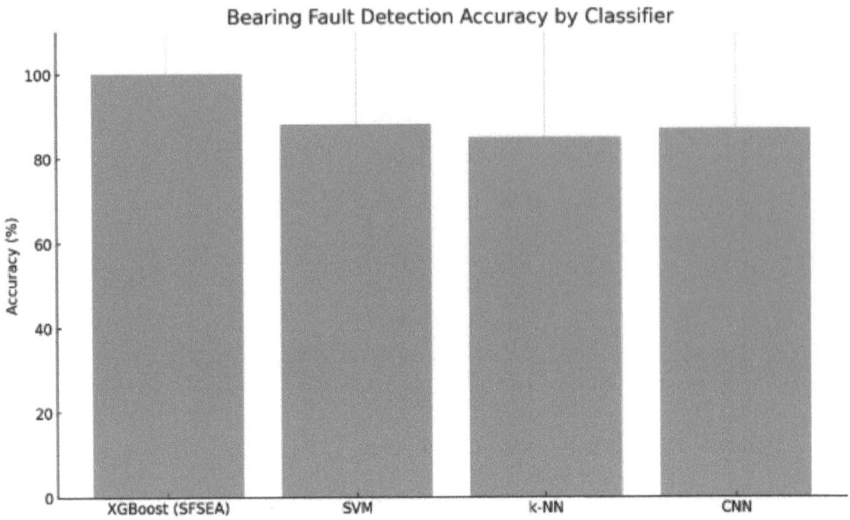

Fig. 9 Accuracy of fault detection by different classifiers

1.15 Impact of Maintenance on Pump Operation

Şevik [11] investigates the impact of maintenance on the performance of high-pressure boiler feed water pumps (HPBFWPs) in power plants. It is concluded from the study that regular servicing leads to better-working pumps, lower repair costs, and higher dependability. The paper demonstrates that proper maintenance boosts the performance of pumps.

Flow rate, pressure, and overall efficiency metrics are examined by the study on the HPBFWP, both before and after applying maintenance. Cleaning out and maintaining the boiler led to a rise of 8% in the flow rate, a pump index ratio (PIR) that fluctuated between 23 and 29%, and a peak valve opening rate of 70%, this valve refers to the control valve (shown in Fig. 10). PIR combines important pump functions such as flow, pressure, and energy to assess the pump's overall performance. Raised valve activity indicates that maintenance helped the performance by clearing obstructions in fluid flow. Most failures of pumps are caused by problems with the stator, rotor, or bearings. While there are more parts that contribute to failure, these three are the most common. Moreover, the study found that AI and condition monitoring help find and remove faulty parts sooner, by relying on well-modeled systems and predictive maintenance techniques.

The study is regarded as useful because it has clear relevance in real plants and shows how efficiency can be significantly raised. Despite this, the report does not include a detailed report of the financial benefits from routine maintenance programs. Although the research looks at AI-based fault detection, it does not suggest how the technology can be used in actual maintenance work.

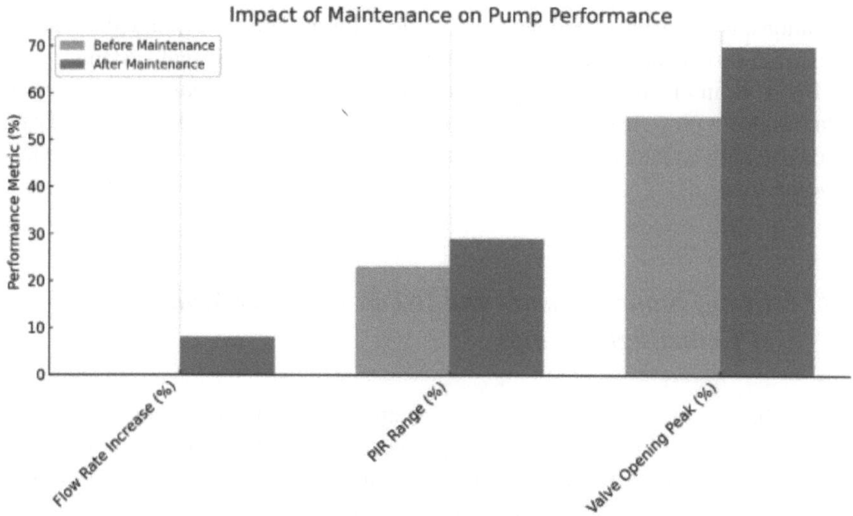

Fig. 10 Efficiency comparison pre and post maintenance

Based on the research, executing maintenance increases pump effectiveness by 8% and allows valves to open faster. This link between valve opening and improved flow shows when there is something wrong with the system. The research paper shows that combinations of AI and live monitoring could help prevent rapid breakdowns.

1.16 Integration of Preventive and Predictive Maintenance Strategies

Yıldız and Soylu [12] present an integration of preventive and predictive maintenance policies via a decision-table framework to improve system sustainability with limited resources The paper focuses on the issues of maintenance planning, examining how to balance efficiency, costs, and staff availability.

The proposed methodology uses Weibull analysis for preventive maintenance scheduling and machine learning algorithms for predictive failure detection. Using the condition of equipment and the resources on hand, it changes maintenance actions to let planners put important jobs ahead of lesser ones. Results from using the method on actual datasets prove that it helps save resources, lower idle time, and prevent unnecessary upkeep. The trade-offs between preventive, predictive, and corrective maintenance are also studied, and it is shown that the decision-table strategy gives a flexible alternative to traditional policies.

A major advantage of this research is the practical and organized tool it provides for making decisions while dealing with real-time challenges. It works best when failure forecasts are precise and the data provided by the system is correct, focusing on sensor operations. Besides these findings, real-world solutions are not discussed, including preparing workers and linking the innovative approach with current systems.

The work in this research supports intelligent maintenance planning by merging data insights with well-structured choices. More work should be done to modify the model for various industries and add adaptive learning methods to boost the model's ongoing success.

1.17 Use of Smart Sensors and IoT to Intensify Monitoring of Centrifugal Pumps

An IoT-based predictive maintenance system for centrifugal pumps using smart sensors is presented by Chen et al. [13]. The study proposes using both wired and wireless sensors for real-time monitoring of vibration and temperature for bearing wear, misalignment, and cavitation fault detection.

It describes the development of a wireless Microelectromechanical System (MEMS) sensor, and a wired sensor integrated with a cloud-based data analysis system. The wired sensors give continuous real-time monitoring, while the wireless sensors communicate via Narrow Band IoT (NB-IoT) and run on a low-power mode. Because these sensors test continuously, their accuracy rate rose to over 85%. Unlike ordinary sensors, wireless sensors can monitor slow weakening effectively as they collect data over extended periods, allowing for gradual trends to be detected easily.

Experimental work with a Grundfos centrifugal pump demonstrates the effectiveness of the study. A limitation is that wireless sensors may not capture short-duration faults such as water hammer events due to intermittent data transmission. Although the system makes use of predictive analytics, it fails to detail the machine learning approaches used to find faults.

In general, this research presents a robust framework for smart pump monitoring, as well as the feasibility of IoT-based predictive maintenance. Future efforts should be aimed at better sampling data from wireless devices and adding AI to fault detection to enhance present predictions.

1.18 Vibration Analysis in Pump Fault Diagnosis

In Mohtar [14], found that vibration analysis together with MATLAB and fast Fourier transform (FFT) can be used to find centrifugal pump faults. The investigation reveals that vibration signals reveal faults like imbalance and misalignment, making it easier to measure the reliability of the pump.

Researchers use an experimental method, recording vibration changes under different operating circumstances. The longitudinal and transverse vibration spectra measured are analyzed, and amplitudes at specific frequencies indicate the presence of faults. Following ISO 10816 and LUDECA's fault diagnosis standards, it was discovered that unbalance and misalignment play major roles in reduced pump performance. Most strikingly, greater vertical vibration amplitude, in the form indicated by the FFT, relates to more severe unbalance problems.

Vibration analysis is shown by the study to play a key role in preventive maintenance and helping prevent overlooked pump breakdowns. Even so, the study has some limits, as the tests ran on a single-speed pump in a laboratory, different from the usual conditions seen in real-world applications. Wavelet transforms might enhance the sensitivity for finding faults by providing an alternative method for signal processing.

It was shown that using vibration analysis is a suitable way to monitor how well a pump is running. As we continue to develop this technology, it is a clever idea to work with more than one speed for the pumps and to connect real-time monitoring with machine learning, so that we can introduce advanced predictive maintenance systems.

1.19 Machine Learning-Based Pump Fault Detection

Menanno and Salsano [15] introduce a machine learning method for predicting faults in centrifugal pumps using signals obtained from vibration and acoustics. This work fits with Industry 4.0, as it focuses on using AI in maintenance to reduce interruptions and optimize the cost of maintenance.

The model used supervised learning and applied a sliding window approach to preprocess the data. Afterward, the model was trained with a support vector machine (SVM) and obtained 99.7% accuracy when classifying five pump conditions such as normal operation, broken impeller, clogged impeller, and bearings with inner or outer race defects. Measurements from high-frequency sensors are used, and the dataset was obtained from the Institute of Electrical and Electronics Engineers. SVM outperforms such classifiers as decision trees and ensemble methods in terms of achieving predictive accuracy.

This study analyzes the same data with acoustic and vibration signals which makes fault detection more reliable and hence adds value. The study does not address real-time deployment challenges or the computational cost of SVM-scale implementation. Although the model does well on average, it is not reliable for detecting the "normal operation" state and needs improvement.

Overall, the study has shown that AI can predict when a pump needs to be serviced. Further work should consider live applications, evaluate costs, and investigate enhanced ensemble methods for better fault classification in industry.

1.20 Analyzing the Growth and Future Innovations in Predictive Maintenance with AI

Ucar et al. [16] present a detailed review of the key components, trustworthiness, and future work of AI-based predictive maintenance (PdM) applications. The research demonstrates that using machine learning, deep learning, explainable AI (XAI), and similar technologies can help maintain the reliability and efficiency of systems.

This paper also makes a major contribution by focusing on explainability and trustworthiness in AI-driven PdM. The study divides AI-PdM into its essential components, namely data acquisition, preprocessing, predictive modeling, and decision-making. It further examines advanced ways such as generative AI, collaborative robots, and the newest digital technologies that can be used for quick updates during maintenance. Moreover, the study highlights several problems with integrating AI, mainly problems with data quality, interpreting models, and relating to ethics.

The study provides an excellent introduction to the theory, but it does not explain how the framework could work in practice. The paper considers AI ethics and transparency, but it does not propose solutions for their application in industry. While blockchain and federated learning are discussed, proof of their usefulness in the real world has yet to be established.

The results of this research show a helpful path to improve AI-based predictive maintenance. Further studies should focus on adjusting the model in real-time, overcoming deployment challenges and studying real uses in industries to support practical use.

1.21 Conclusion

According to this review, sensors, data collection, processing signals, and artificial intelligence are the most crucial factors for achieving predictive maintenance of centrifugal pumps and various rotating machines. Sensors detect physical symptoms of machine condition, measuring vibration, acoustic emission, temperature, pressure, current, and inductive oil debris. Feature extraction and machine learning models are used to correlate these signals to failure modes, which allows for early fault detection as well as estimation of remaining useful life (RUL). On the other hand, no sensor type is perfect, as inductive sensors cannot find out about early wear or cracks inside the machine. Therefore, for a robust monitoring system, it is necessary to combine data from more than one type of sensor.

In addition, the data acquisition techniques used play a vital role in the effectiveness of predictive maintenance. Wired high-frequency systems supply a lot of data, though setting them up takes a lot of resources. Unlike other approaches, edge and wireless devices can link too many things and provide more flexibility, though they might not catch very brief problems and can become stuck in some situations.

Sensor synchronization is necessary to correlate signals properly, although it remains difficult but is also important for accurate monitoring.

Several methods are employed with FFT, CEEMDAN, auto-encoders, SVM, XGBoost, and CNNs to analyze sensor readings and discover patterns of failure. Results from tree-based models are easy to understand and come fast, while it takes deep learning more time and computing power to do the same. Dimensionality reduction and noise control methods in SFSEA and wavelet transforms lead to better model accuracy.

With AI, predictive maintenance now stands out for performing automated identification of faults, finding anomalies, and estimating RUL in the most advanced way. Now, these advanced models use explainability (XAI), edge computing and federated learning to address the requirements of real-time, large-scale, and safe industrial implementations. Yet, it is still difficult to apply AI successfully where behavioral changes happen and the data used is inconsistent.

Along with other results, these conclusions will provide the basis for developing a smart predictive maintenance system. The system should be strong, scalable, efficient, make sense of various sensor inputs, collect data well, detect problems, predict when they might happen, and thus improve performance and expenditures.

Table 1 presents the best uses and main drawbacks of popular strategies for detecting problems in centrifugal pumps.

2 System Description and Methodology

2.1 Introduction

The chapter describes the approach to create a predictive maintenance model for centrifugal pumps through artificial intelligence methods. The system applies actual vibration measurements from an Omani oil and gas company and performs intensive data cleanup followed by elaborate feature development leading to neural network-based model development. The goal is to develop a stable predictive model which forecasts the useful life span of pumps by providing an estimate of remaining days till failure. This study developed methodological steps that reach both industrial and academic standards by following best practices across machine learning and predictive maintenance systems and data analysis fields. A flow diagram is presented in Fig. 11 for visualization of the overall methodology.

2.2 Data Acquisition

Data for this research was obtained from a leading oil and gas company with extensive operating experience in rotating machinery. Vibration data were obtained of six

Table 1 Comparison of monitoring techniques for pump fault detection

Technique	Detectable Faults	Limitations
Acoustic emission (AE)	Incipient cavitation, turbulence, structural cracks	Susceptible to ambient noise; therefore, may not operate well under a changing operational environment
Motor current signature analysis (MCSA)	Cavitation, rotor faults, electrical anomalies	It is challenging to draw a clear demarcation between cavitation and other motor maladies
Vibration analysis	Unbalance; misalignment; bearing faults; mechanical looseness	Limited sensitivity to incipient faults; noise-prone; requires sensor placement
Thermal monitoring	Bearing wear, friction-related overheating	Slow response time; inadequate for the detection of mechanical failures unless heat accumulates
Inductive wear debris monitoring	Wear and spalling, metal particle generation in lubricants	Cannot detect faults which do not create debris (e.g., internal cracks or onset of fatigue)
Smart sensors (temp, vib, pressure)	Cavitation, misalignment and overheating and pressure surges	Wireless sensors may fail to detect short-duration events; signal sampling strategy varies in accuracy differences
CEEMDAN with MCSA	The beginning of cavitation, developing cavitation based on time–frequency features	High computation cost, incipient cavitation needs accurate calibration for different pump set up
Explainable AI (XAI)	Anomaly detection, RUL prediction, component degradation	Arbitrary labeling of available data will not easily yield useful predictions in anomaly prediction over time. Hard to put to practice in dynamic environments without adjustments
IoT and edge computing	Real-time fault identification (e.g., motors, shafts, rotors)	Integration difficulties; limited by power from the device and its processing ability
Fast Fourier transform (FFT)	Cavitation frequency band detection, imbalance, misalignment	Weak time resolution; not good for non-stationary transient signals
SFSEA with IPA	Minors bearing fault through a harmonic signature	Only validated in laboratory-scale tests; requires further validation under the real-world industrial load variation states

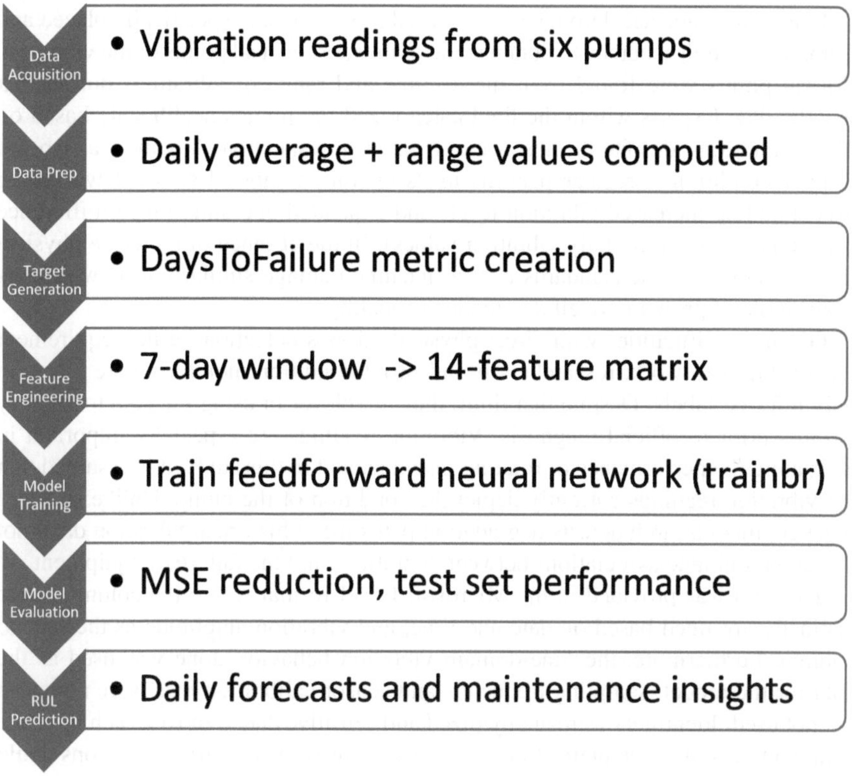

Fig. 11 Methodology flow diagram

centrifugal pumps in a defined operational period. Four out of the six pumps were in failing health and characterized by different modes of failure, while the third was healthy and operated under normal conditions. The four pumps were categorized as unhealthy by professionals in the field due to excessive vibrations. The classification of pumps as healthy or unhealthy was done only through vibration analysis, with no disassembly or physical visit required. Reliability engineers within the company stated that extended, elevated daily vibration levels in both pumps were leading to deterioration and departure from their initial state considered healthy.

Determining the correct location to place sensors needed evaluation. Our industry partner set up vibration sensors at each pump's non-drive end location. Industry practice and dataset data indicate that the sensors were put externally on the bearing housing of NDE to have better access to vibrations there. This study focused exclusively on the NDE. Our choice to use NDE vibration readings was driven by good data availability and data quality across the multiple pumps in the X direction. Expanding the study boundaries to include both NDE and casing data would create better models but needs more time than this project has.

Industrial-grade accelerometers were used as sensors and specifically placed and calibrated to high accuracy to obtain some of the most critical points of the vibration. Critical points were found from the average and range of vibrations during that specific day. Experts within the field categorized the pumps health state based on pump vibrations, despite the absence of a physical inspection. A pump was labeled in poor condition if regular patterns in its operation showed signs of worsening over time like increased vibration levels and expanded operating range (difference between max and min daily vibration values). In the absence of a precise physical failure analysis, these standards reflect machine damage similarly to how doctors analyze blood pressure to tell if someone is healthy.

Given the difficulties with direct physical failures detection, or the requirement for machines to work, researchers in industrial condition monitoring are left with only inferred labels. Despite that limit, they are clever in using the data to identify hazards prior to official diagnosis. Vibration readings are especially important in evaluating the performance of rotating machine, and in this study, it is assumed that the vibration readings currently depict the condition of the pump. Unlike standard practice, this research detects degradation paths using historical vibration data, not through unreliable associations between vibrations and the failure of equipment.

The data was provided in the Microsoft Excel format where the columns were distinctly arranged based on date and time, and vibration amplitude as the second column. Furthermore, the time-domain vibration behavior alone was used as the data for the model; sampling frequency and sensor calibration logs were reviewed but not used. Identifiers were anonymized and sensitive data removed such that data confidentiality was maintained and ethical standards of industry regulations could be complied with.

2.3 Data Preparation

It was MATLAB script DataPrep.m that had the objective of processing the raw time series data at high frequency to make it as streamlined and structured as possible for efficient use by machine learning. The association between mechanical failure and vibration signals is complex, characterized by nonlinearity, hidden subtlety, and inconsistency concerning the timing of events. The fact that there are not always simple and straight relationships between the vibrations and signal failures makes machine learning a suitable approach, as it can model these patterns straight from the collected data. Toward the end of this phase, the dataset was imported and prepared by extracting needed NDE X-axis vibration data. For the sake of arranging data by time, timestamps in the given dataset were converted to MATLAB's date time format, which simplifies the processing of dates and times, and enables in-depth temporal analysis.

Based on the daily aggregation of the vibration signals, there were two important statistics calculated each day: the mean vibration signal, and day's maximum to minimum range. The range shows the readings variance or fluctuations for the pump while the average represents an indication of the central performance of the pump or the selected unit. Daily vibration statistics were logged in an organized table dailyStats keyed by date. When searching for other statistical indicators such as variance and standard deviation for the aggregation of data, neither was incorporated choosing this technique to reduce sensitivity to outliers and increase sensitivity to mechanical disturbance.

The DaysToFailure metric in turn is found to be vitally important in this study since it becomes a substitute for the RUL estimation when the failure time stamps for the supervised learning cases are not available. There are two fundamental assumptions upon which the design of this metric is based. Those discrepancies in vibration measures are regarded as expressions of negative change in performance of pumps. Second, when operated, a process of incremental deterioration of the performance of the pumps occurs, as it should, and indicators of this wear may not occur immediately. The model reveals a failure timeline with gradual deterioration of the pump performance over the extended period without any sense of disintegration triggers. The output produced by the model portrays the probability of a failure over other events; but not the exact date of failure. Engineering understanding which presupposes the association of pump wear and defects and substantial fluctuations of vibration differences over time is the foundation of this method. Some findings may question whether this assumption is valid.

Normal operation was extracted using the average amplitudes of vibration data at the beginning of the trial when the pump was in a good state of service. Variation from stable operation is equivalent to the difference in vibration data for every day and the baseline. A pump did not fail when there is deviation from a regular operation. Subject to the settings alterations, the pump can continue to operate without causing hassles. The deviation from baseline was weighted with a penalty the factor, $k = 0.5$, to get more meaning. At the same time, the age of the equipment was determined by counting days of when monitoring commenced. The final formulation was:

$$\text{DaysToFailure} = \max\left(0.365 - \text{elapsed} - k * \left|\text{avg}_{\text{current}} - \text{baseline}\right|\right)$$

- Max(0, …): Ensures DaysToFailure remains a positive integer
- 365: A theoretical maximum lifespan in days
- Elapsed: Number of days since monitoring began
- $\text{Avg}_{\text{current}}$: Current days average vibration amplitude
- Baseline: Average vibration on Day 1 of the healthy pump
- $|\text{avg}_{\text{current}} - \text{baseline}|$: Absolute deviation from healthy baseline
- *K*: Penalty factor (chosen at 0.5)

The "DaystoFailure" method presented in the chapter is not about actual days, but it was designed to roughly estimate how much longer a component can function based on gathered information. In most instances, the monitoring period showed the pumps were still running well, even though no real failures were found. Failure here refers to when a pump operates poorly, showing signs of excessive shaking, lack of efficiency, or danger to the equipment downstream. As a result, we do not have a reliable day that we know for sure is the "failure day" of a healthy pump. So, the model includes both the number of days and the change in vibration pattern (standard deviation) to show a decline in health. Even though we are not measuring variability in time periods, it helps notice potential problems at an early stage. MAE is calculated after the system has been in operation, detecting, and logging incidents where the performance is below thresholds. Because of this, using the model's forecast to schedule maintenance lowers the chance of being surprised by downtime due to equipment failure.

We applied two separate strategies to track the course of pump degradation during the solution development period. As equipment operates continuously for longer, pump performance deteriorates, and it becomes more sensitive to vibrations. DaysToFailure is produced by combining operational lifespan and fluctuations in expected vibration patterns. The algorithms of machine learning were applied to identify even minor changes in the performance indicative of likely problems with the pump mechanism.

To improve clarity in our dataset, vibration ranges were built into the model training process while maintaining them in the target variable. Fully documenting past failures of equipment was crucial for advancing our predictions, and adding vibrational range data helped to understand more about the model's workings. Our approach allowed the model to determine gradual and drastic changes in vibration levels and, subsequently, to accurately detect potential problems with the pump.

The detailed MATLAB script is provided in Appendix 1.

2.4 Feature Engineering

A data structure was built, which was appropriate for machine learning reflecting the format presented in BuildFeatures.m script. To synthesize each sample, vibration data, recorded seven days before each observation, was picked up and utilized. We built a feature matrix vibrating with 14 numerical values where each pair shows the average and range of the previous seven days vibration data.

In the present example, a vector is said to be a set of numerical values that indicate a physical direction differently. Feature vectors are utilized to summarize the history of vibration of the pump with baseline statistics such as mean and signal ranges.

The choice between data from seven days ago or today's data was backed by a mix of technical insight and actual testing with current world data. A too narrow or too broad interval did not produce positive results—short intervals could not warn against future problems and extended intervals blurred accuracy due to outdated information. Since no previous studies proved this time, our experiments demonstrated that a 7-day window provided the highest accuracy compared to all other criteria assessed.

Using this approach, the model ensured that the most recent temporal changes were mastered to be able to distinguish normal variations from progressive deformities. In predictive and slow decline systems meanings of temporal embedding are essential. For each of the feature vectors, we predicted DaysToFailure label for next day relative to window of 7 days. The problem, therefore, was transformed into a temporal sequence prediction task (or univariate regression, or time series forecasting) and hence was perfectly poised to assess the neural network-based models.

To allow for strong evaluation, the data was divided into three subsets: training, validation, and testing. Manual splitting was performed to preserve temporal order and avoid data leaks resulting from dividing time-dependent values between training and testing when they are close in time. An 80% of the samples were allocated to training, another 20% to validation with any leaked data reserved for testing. The partitioning provided for model performance estimation using unseen data, which will ensure an unbiased estimate within the model's generalization performance.

See Appendix 2 for the feature engineering script.

2.5 Model Development and Training

TrainEvaluate.m script was used to design and train the predictive model. As a result of the nonlinearity of the regression problem, a decision was made that a feedforward neural network would be useful. A design of the form with one hidden layer containing ten neurons was selected after several tries considering both performance and computational efficiency.

Since it is widely accepted for small and medium-sized datasets, 10 hidden neurons were chosen to begin with. In this case, where the objective is to make a continuous prediction with a simple regression model, 10 neurons are likely to give the best trade-off between learning and the danger of overfitting. It is common to use 10 when testing models even though 5 or 15 were not assessed, especially when data is not abundant and only a single hidden layer is present.

To mitigate the threat of overfitting, and to facilitate rather better generalization, the trainbr algorithm was used to train the model. Bayesian regularization was implemented because it improves model efficiency by minimizing a cost function that consists of the mean squared error and the sum of squared weights. As this strategy is used, the model becomes simplified, and high weights are opposed.

The functioning of neural networks is dependent on neurons sending signals over set weights. Signals that traverse through the network are influenced by weights that regulate not only the direction in which information goes but also its shape. Bayesian regularization forces the model to take a simpler transformation, thus resolving overfitting and providing both reliable and flexible results.

Through the mapstd function, the input and output data were normalized using a transformation that left each variable with zero mean and unit variance. Lack of standardization of inputs disrupts the convergence and destabilizes the updates of weights in the neural networks making the convergence random. Instead of using standard data partitioning, the network used custom index values created by the previous script. Through this approach, it was possible to have exact separation of training, validation, and test sets as well as retain the time order of the data, thus keeping results from various runs on different models comparable.

The network shown in Fig. 12 uses 14 input values that represent vibration readings for 7 days, 10 neurons in the hidden layer, and an output neuron to predict the number of days before the vibration sensor fails. In every layer, W represents the weights and b denotes the bias. The main function of the long blue rectangles is to change the computed sums—linearly in regression output and nonlinearly for the hidden layer.

The weight and bias values are what the neural network picks up as it is trained. Weights decide how much one neuron will affect another and impact the outcome of the neural network. With biases added to their inputs, models can reach and keep a fitting function that is less likely to draw short of the data. During training, the tool adjusts the weights and biases to minimize any error in its predictions, allowing it to notice the underlying parts of the vibration data that reveal how the pump has diminished.

Training and evaluation procedure of the neural network model is outlined in Appendix 3.

2.6 Model Evaluation and Optimization

The first estimates of the neural network produced an MSE value that was roughly 1400 and undesirable by our standards. MSE is a dimensionless measure that is an average of squared error. MSE is the mean of squared differences between model's predicted (RUL) and actual ones (DaysToFailure) and the lower the score, the more accurate the model is. At such a prominent level of MSE, the forecasts of the model may not be dependable and may confuse the early signals for effective planning of maintenance. Therefore, systematic optimization was necessary. Various optimization techniques were used; features' usage was reconsidered, the most important settings of the model were changed, the overall structure was tweaked to decrease the risk of the overfitting.

There was improved model representation capacity by including the daily average and range values. Though not so sophisticated, however, these attributes include essential mechanical considerations such as signal mean and volatility which provide an important early alarm to eventual breakdown. To avoid overfitting when learning the data, a regularization parameter is introduced, which would lightly simplify the model to minimize its dependability on the training set. With the original regularizing parameter set at 0.1 where trade-off between complexity and error could be made, we proceeded to set the optimization routine. To maintain the capacity of the model, but not memorize the training data excessively, the number of neurons in the hidden layer was set to 10.

Using the "max_fail" approach, we made sure the training would stop when the validation performance would plateau. This technique is applied according to the established standards in neural network research. Our cross-validation approach paused further training when model performance on validation data reaches stability.

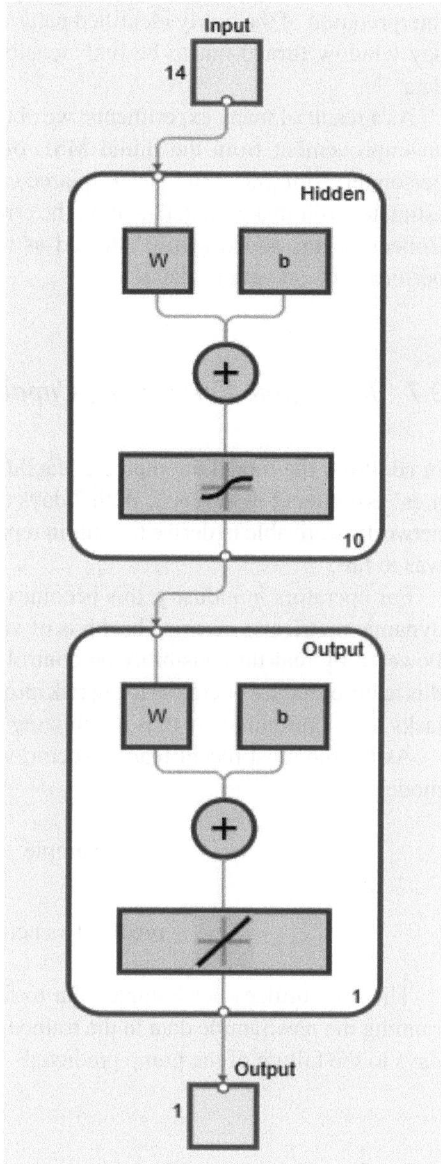

Fig. 12 Architecture of neural network

Application of visualized data made it possible for us not only to evaluate the convergence of the model but also to identify the signs of overfitting. It was important to track both the training and validation loss over epochs to see how the model performed and check for signs of overfitting. We further investigated the impact of different window sizes—from fewer than seven to more than seven days. The context was reduced within the shorter windows, and the longer windows impeded interpretation of the newly identified patterns. Consequently, the choice of the seven-day window turned out to be both sensible and in accordance with experimental data.

As a result of many experiments, we obtained a final test-set MSE of about 300—an improvement from the initial MSE of 1400 which is useful for maintenance personnel. Along with the mean squared error, the mean absolute error (MAE) was estimated as it makes understanding the error in days more interpretable for the user. Collecting this added metric allowed us to check the clarity of the model and its performance on various datasets.

2.7 Real-Time Forecasting Capability

In addition, the model attempted to facilitate real-time predictions and future failures' assessment in advance. With 7 days of recorded vibration as the input, trained networks were able to derive a result in representing days until an item of equipment was to fail.

For operators in industry, this becomes a valuable component because access to dynamic metrics of machine health is of vital importance in planning interventions. Powered by real-time visibility on control room dashboards and mobile platforms, this feature enables operators to put risk into context briefly and schedule maintenance tasks in an optimum way thus minimizing downtime.

As for the most recent feature vector, we used the script line to pass data to the model:

$$\text{newSample} = X(:, \text{end});$$

$$\text{predLast} = \text{net(newSample)};$$

This line utilizes newSample data to feed it to net and create a prediction. By running the newSample data in the trained network, we get predLast, the number of days to the failure of the pump predicted.

2.8 Computational Environment

All analysis was done in MATLAB R2024b. Simulations were conducted in a system running on Intel Core i5, 32 GB of RAM and solid-state drive for file handling improvement. The effective model development was made possible due to the availability of well-put functions for designing, training, and analyzing network performance through the neural network toolbox.

With computational needs of the model being relatively efficient (several seconds in many cases), one could conduct many experiments in close succession, which would result in multiple improvements. The scripts were also modular by nature; hence, they made versioning and comparative analysis on so many different configurations quite simple.

2.9 Ethical Considerations

Real industrial data utilized introduced an ethical obligation that no identifiable or sensitive operation data were exposed. Before analysis, all datasets were anonymized. Company names and any time stamps linked to operational events were either eliminated or anonymized.

Apart from confidentiality, another ethical aspect related to safety was also discussed. The pumps received no artificial stress, and their natural operation was not changed for this experiment. The synthetic target variable, "DaysToFailure," is a simulated estimation from vibration deviation that was not due to forced degradation. With this design, no equipment was put at risk in the name of academic experimentation.

2.10 Limitations and Future Work

The ability of the model to estimate pump failure timing is remarkable, but some limitations remain that require further scrutiny. First, the model was developed using X-axis vibrations data obtained at non-drive end as an autonomous input source. Expanding the input space of the model to include vibrations at the non-drive-end and all axes will, in future model accuracy, be anticipated to yield high gains.

Also, the model's effectiveness may be compromised as it was built based on data from only six pumps covering various levels of health. Inclusion of information from various pump designs, factories, different workload situations, further context may result in development of a more generalized and universally applicable model.

The current setup analyzes vibration samples from a window separately and does not consider features that show up in the frequency domain, for example, issues with bearings or impellers. Future improvements might include exploiting frequency-based features using techniques like fast Fourier transform or spectrogram analysis or wavelet decomposition to detect small abnormalities in the time series data.

The model is not intended to facilitate real-time condition monitoring at the present time. The supervisory control and data acquisition (SCADA) or Industrial Internet of Things (IIoT) connections for the real-time data streaming and the visualization of the predictive results will be the probable future directions for the model. Important to success factors include minimization of processor delays, system reliability, and the implementation of effective threshold-based alerts.

Since neural networks are powerful and adaptable when dealing with nonlinear data, they were chosen for this analysis. Because they are adjustable, they can be used for many kinds of predictive maintenance. On the other hand, such as ensemble methods, support vector machines, or recurrent neural networks, could also obtain results that are at least as good as or better than the results obtained by logistic regression. How well feature engineering is done and what the dataset is like are strong factors in this decision.

Looking for ways to refine the models and introduce the element of interpretability would contribute to making people more welcoming and confident to use them and make maintenance personnel understand the predictions and act in response.

Since these suggested improvements are flexible, the existing framework can be considered a firm foundation toward advanced and real-time predictive maintenance systems.

2.11 Summary

In this chapter, the stepwise procedure of a methodology to develop an AI-driven predictive maintenance system with vibration data from centrifugal pumps has been explained. Among others, the required period for conclusion consisted of rigorous data preprocessing, carefully designed feature engineering based on time window statistics and neural network training with the strong focus on regularization and generalization. The optimization steps were done step by step, starting from the initial value of MSE 1400–300 to ensure that the model is ready for the real-time usage. Our team used MATLAB software and its neural network toolbox tools to develop our system through experiments.

2.12 Insights from Literature Review

Insights from the literature review guided the designing of many parts of the methodology. The choice to use vibration data for pump health monitoring rests on work by groups such as Bruinsma et al. [6] which proved its usefulness in diagnostics. Developing the DaysToFailure measure was motivated by Mohammed's [7] findings which highlighted how surrogate targets can substitute for unknown failure times using regression methods. Similarly, SFSEA authors [10] and many others from vibration-based research apply statistics over time windows, like our 7-day one, rather than showing and studying the raw signals. Because several AI-based PdM studies (such as by Rajaoarisoa et al. [8]) mentioned the importance of avoiding overfitting while still being able to generalize, a Bayesian-based regularization system was introduced.

3 Results and Discussion

3.1 Introduction

This chapter discusses the results of the predictive model that was created to predict the remaining useful life (RUL) of centrifugal pumps from the vibration data. Neural network was trained and tested on sets of vibration data that reflected the operational conditions of pumps, healthy, and malfunctioning, recovering after maintenance. The analysis shows both the power of the predictions and the usefulness of the model in real-life operations.

3.2 Neural Network Construction and Training Productivity

The single-layer feedforward neural network predicting architecture applied in all datasets had 14 input features, 10 neurons in the hidden layer, and 1 output neuron representing the predicted no. of days to failure. The hidden layer consists of 10 neurons, each connected fully with all 14 input features, so that each could train on each input independently with weights and biases before passing activation. Bayesian regularization (trainbr) training algorithm was chosen as it was especially appropriate for this application because it could deal with noisy data and avoid overfitting without the need for an additional validation set. In all of them, however, the model converged at some point when the early stopping conditions were met, particularly by crossing the upper boundary for the adaptation parameter (mu), which is usually after <1000 epochs. The training progression was similar in all conditions, and final performance was measured via values of test-set mean squared error (MSE).

The training of the neural network is shown in Fig. 13 with the mean squared error gradually decreased to best value of 95 in 35 epochs. Regardless of setting it up for 1000 epochs, the training stopped due to convergence criteria set by the Bayesian regularization algorithm. It is confirmed in Fig. 14, which shows constant decline in gradient and growing regularization parameter (Mu) and not validations triggered. If the model does not cut down error as it should, Mu goes up, and if it grows above a specific limit, training comes to a stop, as that means the model has converged. The slope of the performance function is what the gradient indicates, and it falls when the model is approaching a helpful solution.

A regularization value of 0.1 was chosen as the model trained to make sure it did not become too specialized and developed simpler solutions. Values of the regression coefficient R shown in the plots should be close to 1 for there to be a strong relationship between predicted and actual data. In Bayesian regularization, Mu limits the size of the learning steps and the training finishes when it meets a specific threshold.

The regression plots in Fig. 15 indicate predictive strength of the model which has R-value of 0.95175. How accurate the model fits the expected values is determined by the regression coefficient R; the value of R should be remarkably close to 1 for

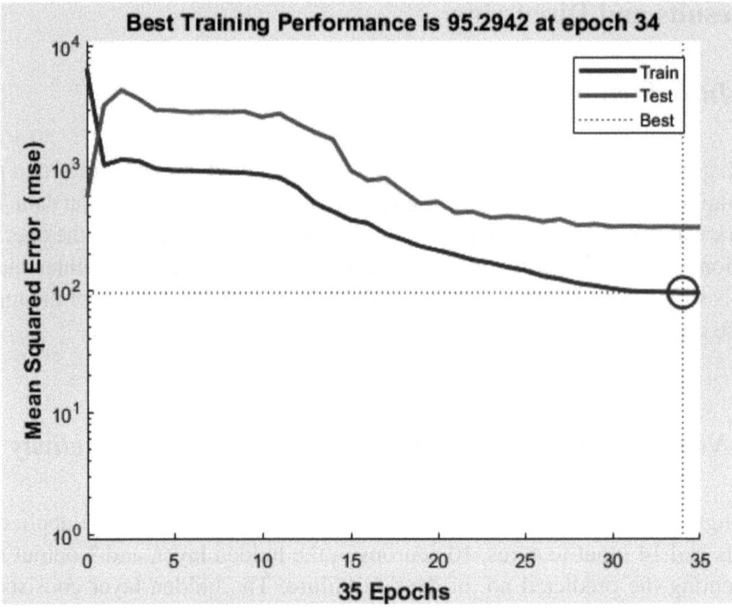

Fig. 13 Trend of MSE across 35 training epochs

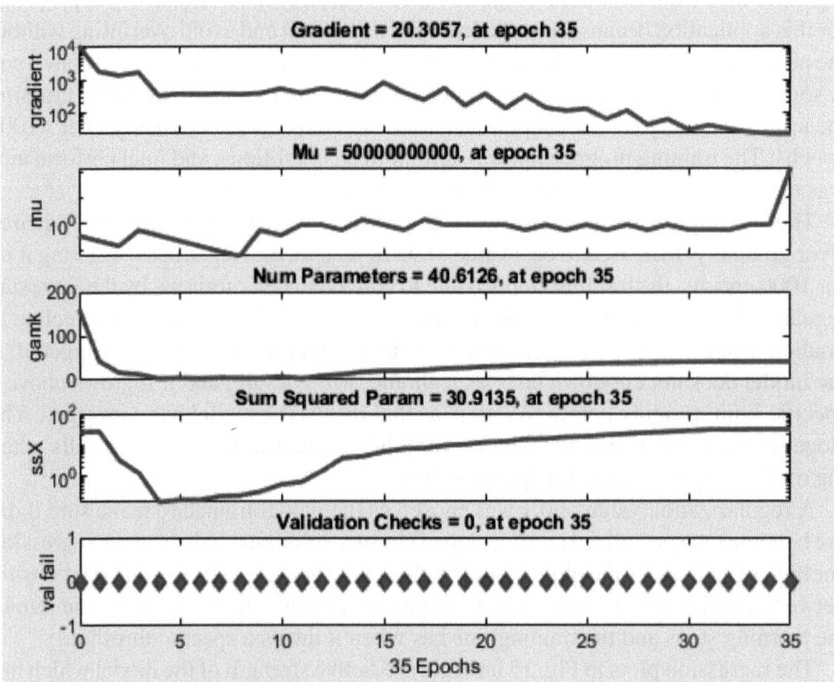

Fig. 14 Training state showing gradient, mu, parameter count, and validation checks

top accuracy. Each separate straight line on the regression plot gives the level of correlation for training, validation, and test sets. The data from the test set was not used for *R*-scoring, and this may be evidence of an issue with data partitioning.

Although the *R*-value for the neural network was 0.95, other SVM models reached as high as 99.7% classification accuracy as applied by Menanno and Salsano [15]. Nevertheless, since their goal was fault detection, it was not easy to compare them to work done in RUL estimation. All the same, it underlines how effective machine learning approaches can be for checking pump performance. Finally, the error histogram given on Fig. 16 indicates rather a symmetrical error distribution with zero as its center, which verifies the robustness of the predictions of this model during the training process.

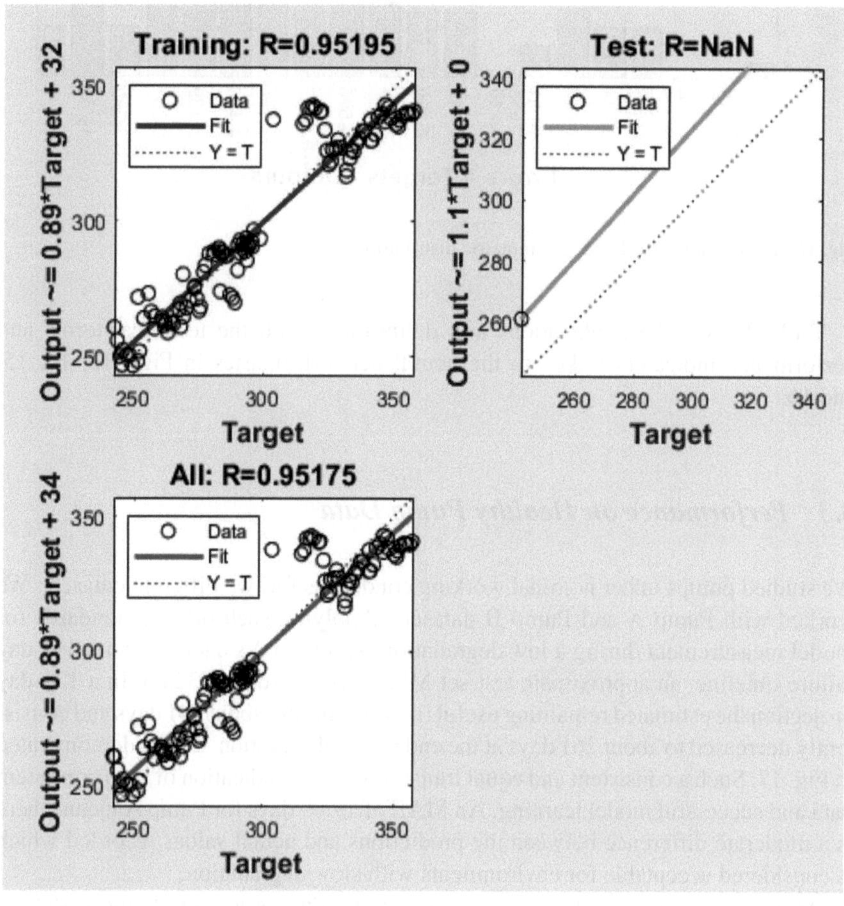

Fig. 15 Regression plots between predicted and target values

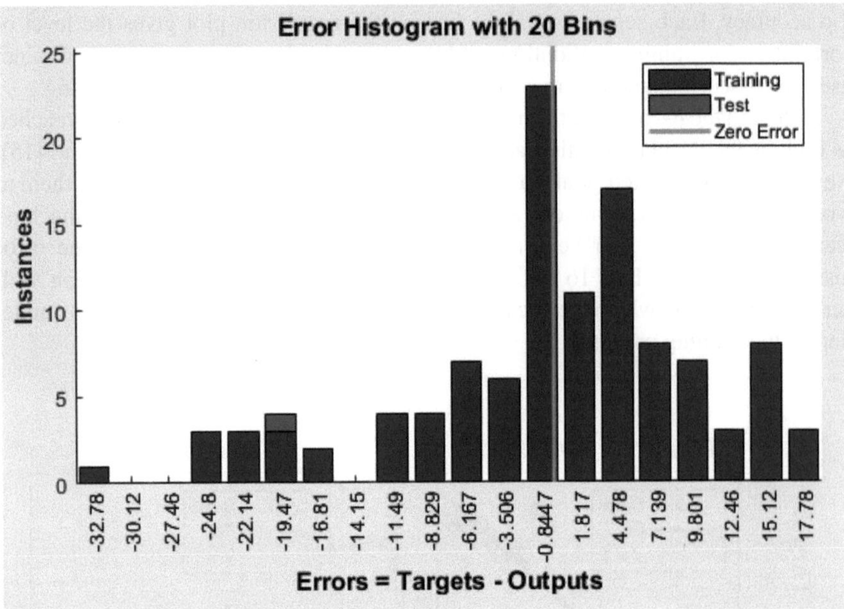

Fig. 16 Error histogram showing symmetric distribution

Table 2 gives the explanations and definitions of all the technical terms and performance indicators linked to the neural network metrics in Figs. 13, 14, 15, and 16.

3.3 Performance on Healthy Pump Data

We studied pumps under nominal working conditions for two different datasets: We worked with Pump A and Pump B datasets. Applying such offered standards for model measurement during a low degradation. At Pump A's dataset with a 365-day failure timeline, an approximate test-set MSE was recorded at 325.6. In a 121-day projection the estimated remaining useful life was initially about 341 days and consistently decreased to about 261 days at the end of the observation. This is demonstrated in Fig. 17. Such a consistent and equal transition was an indication of both consistent data and successful model learning. An MAE of 19.09 days for Pump A means there is a moderate difference between the predictions and actual values recorded which is considered acceptable for environments with slow degradation.

Although both operated within a common 365-day period, the Pump B dataset was found to be more dependable with an observable lower test-set MSE of 9.2. The predicted RUL values ranged narrowly between 352 and 305 days during the 38-day interval, as shown in Fig. 18. The extraordinarily low rate of error indicates high signal

Table 2 Explanation of neural network training metrics from Figs. 13, 14, 15, and 16

Term	Explanation
Gradient	The slope of the curve; showing how quickly the model is learning
Mu	Mu (momentum term) controls the rate of adaptive learning; lower values indicate stable convergence
ssX	The sum of squares of the input data (*X*); used for normalization and calculating error
ssY	The sum of squares of the output data (*Y*); similarly used in loss or error function calculations
Val_fail	The number of consecutive times where the validation performance has not improved; will trigger early (or stop) if it exceeds max_fail
Epoch	A complete cycle through the training dataset during training
Performance	The loss value (MSE) computed during the training, validation, and testing phase
Training state	Shows the training progress, which includes performance, gradient, and other metrics over epochs
Validation	Represents the performance of the model on the validation dataset
Test	Represents the performance of the model on the test dataset
R (correlation coefficient)	Represents how well the predicted output matches the actual target values. Values closer to 1 show a strong correlation between model prediction and true values
Bins	Intervals of grouped similar error values within the histogram

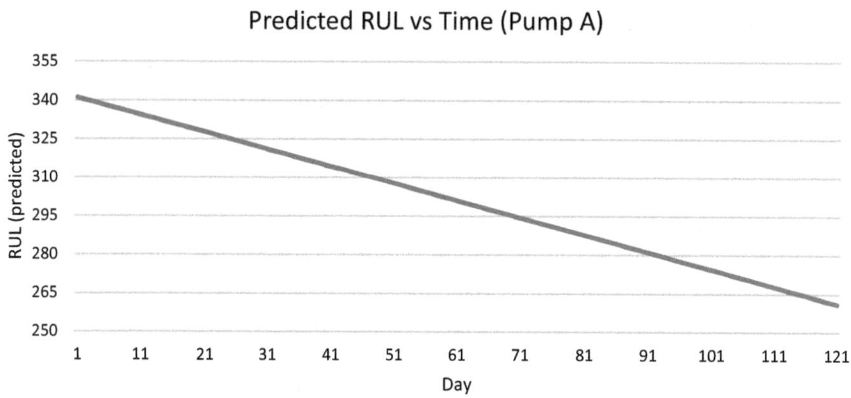

Fig. 17 RUL prediction trend for pump A

reliability and confirms the ability of the model to provide a trustworthy description of slow degradation in sound machines. The small MAE value of 3.29 days proves that the model predicts operational status well and can be used for healthy pump decision-making. Test results from both sets of data validated the model's ability

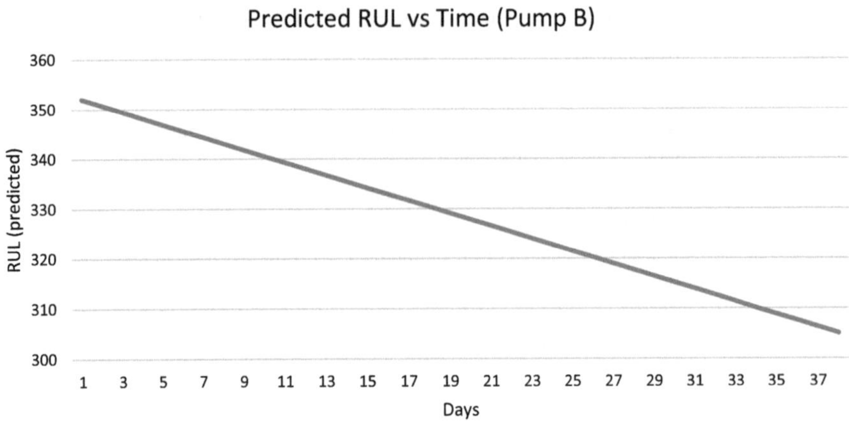

Fig. 18 RUL prediction trend for pump B

to identify early pump deterioration without triggering unnecessary false alarms or underestimating the rate at which pumping has aged.

Note: For the purposes of this analysis, the endpoint for RUL is regarded as the day upon which vibration amplitude exceeds a threshold that is suggested by authorities as being indicative of failure.

3.4 Results from Unhealthy Pump Operation

From the Pump C dataset shown, the behavior of a pump during constant degradation with no maintenance intervention was demonstrated. This set also scaled to 365 days showed a more substantial test MSE of 475. Values of the predicted RUL went down gradually over 121 days of operation from 347 to ~248 days, which is presented in Fig. 19. This shows that in 121 days of operation, the remaining life goes down by 99 days, which indicates a slight improvement in performance. While there was more variability than with healthy data, the overall trend was still consistent, with steady decline. The nonlinearity of a failing pump manifests in nonlinear dynamics, and, as a result, the vibration patterns are nonlinearly changing and, therefore, the predicted values are fluctuating. Although the increased error value, the model still had the capability to register the downward trend, thus signifying the beginning and the emergence of failure over time. An MAE of 22.98 days shows that the prediction remains dependable, despite higher levels of noise in the signal due to degradation.

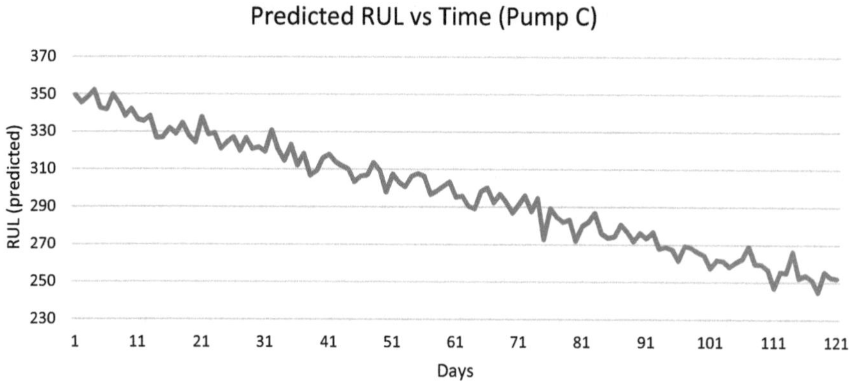

Fig. 19 RUL prediction trend for pump C

3.5 Failure Imminence and Maintenance Events

We evaluated the model's capacity to solve real-life problems in pump health monitoring, including surprise maintenance and system performance gaps by using the datasets like Pump D and Pump E. Restricting the datasets to 90 days allowed closer analysis of changes in remaining useful life (RUL) after faults or maintenance which enabled us to perform short-term trend analysis.

During a 121-day observation period, Pump D had two maintenance events: minor and major. The algorithm showed constant RUL decline which resulted in reaching zero between Day 87 and Day 100, this data is shown in Fig. 20. Despite this, the model's prediction errors were large: an MSE of 1463 and an MAE of 40.33 days which shows a lack of precision during complex failure development and the presence of noisy measurements. The main reason for this high error is that there were uneven vibration spikes around Days 75–85 and no clear maintenance labels were provided in the readings. Because these events were not shown as errors, the model struggled to figure out if something was somewhat recovered or truly failed.

Zero RUL prediction made by the model implies that the model identified the pump as failed or close to failure, consistent with the normal minor repair phase. RUL at day 120 spiked to 38.7 days after intensive maintenance efforts that aimed at getting the pump back in working condition. One of the strengths of the model is the ability to adjust its evaluations quickly with changes in operations irrespective of information about the maintenance schedules.

For Pump E, RUL was at zero and stayed on zero from day 80 to day 120 indicating an enlarged critical period. The RUL on Day 121 increased to 36.3 days, which meant that the intervention was successful. The graph in Fig. 21 is used to demonstrate this. The experts' detailed analysis on vibration data most probably would put a right intervention at a right time. The model showed its capability to recognize and observe the recovery of system's normal operation after the critical period. With MSE of 763 and MAE of 29.13 days, the model showed decent ability to predict the degradation

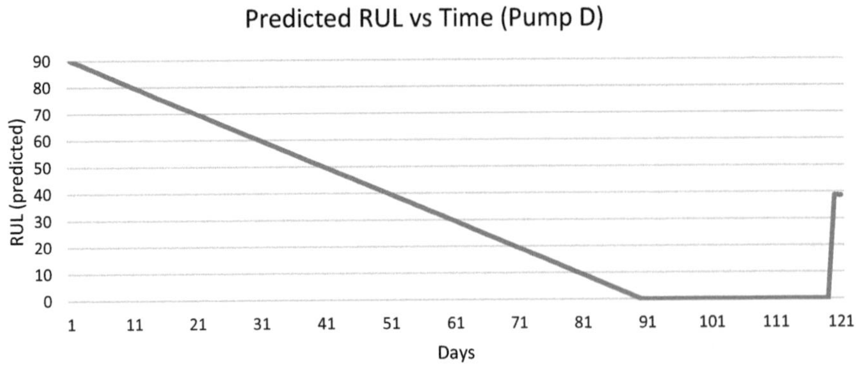

Fig. 20 RUL prediction trend for pump D

phase as well as the recovery from maintenance. Details of the intervention have not been explained, but the experts tell us that the pumps were extensively repaired.

The accurate evaluation of these significant changes in RUL is also only possible with access to appropriate contextual data such as maintenance event records. When the data is not available properly, such deviations can be misclassified as anomaly and overfitting results. However, the fact that the model can identify the restoration of system health evidence a major potential for application of the model in predictive maintenance scenarios.

A major result is that the model learned to predict a system's recovery, even though no data was supplied about the timing of the repairs. Such behavior demonstrates a prominent level of generalization since the neural network learned not only about gradual degradation but also about sharp improvement of the performance solely from the variation of vibration patterns. The tacit detection of recovery patterns

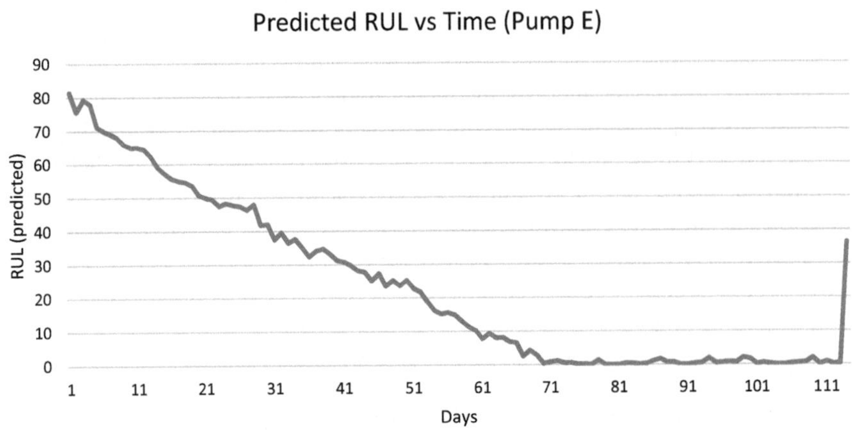

Fig. 21 RUL prediction trend for pump E

including the unexpected spike of RUL after maintenance confirms the ability of the model to respond effectively to field operation variations without specific supervision on annotated intervention points. This adaptability is important for application in dynamic industrial settings, where labeled maintenance logs are usually incomplete or non-existent.

3.6 Short-Term Degradation and Intervention Recovery

An important part of our research was based on a specialized dataset, called Pump F, which deals with unbalance issues in pumps. This dataset covers a period of 40 days and was examined utilizing a RUL model for 90 days. At this point, the model showed that RUL would decrease from an estimate of 75 days to about 22.9 days by the 39th day, as displayed in Fig. 22. 469 and 22.84 days for the MSE and MAE, respectively, mean that the model kept a strong ability to forecast RUL data even as the degradation was not equal over time. The model reacted swiftly when maintenance was conducted, making the predicted RUL rise to 62.0 days on Day 40.

Training data did not include Pump F which was reserved for testing how the model worked in unknown situations. Nevertheless, the model captured the decay of components and responded well to maintenance, showing it can be used on various kinds of pumps and in unknown conditions.

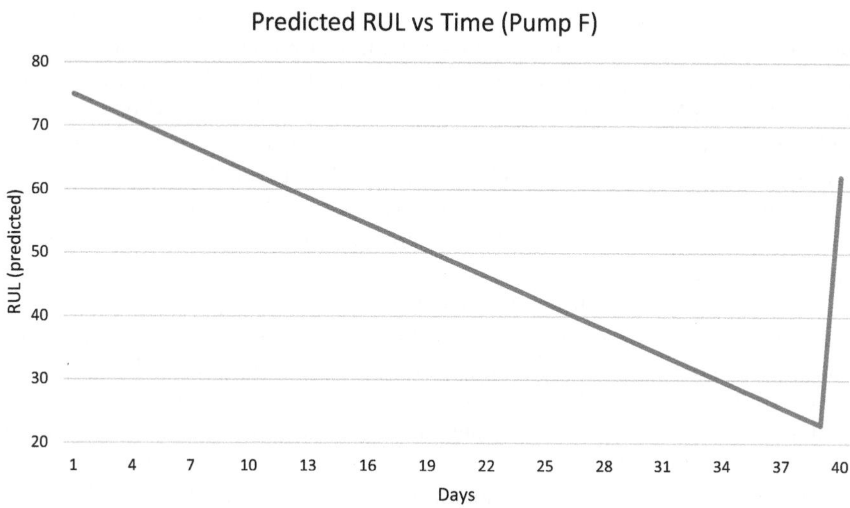

Fig. 22 RUL prediction trend for pump F

3.7 Observing Maintenance Through Predictions

From faulty or problem pump datasets, a clear trend appeared in which maintenance events yielded short- to medium-term improvements to the predicted RUL. In Pump D, the prediction only spiked after significant maintenance action was done. The same applied to the incident in Pump E. These cases demonstrate the possibilities that AI diagnostics can demonstrate prompt feedback on the performance of maintenance work. For example, repeated or worsening post-maintenance predictions may mean incomplete corrections or improper detection of the cause. On the contrary, a rebound in RUL provides a proxy confirmation that the required corrective action did work.

These findings also point to a major problem: the model does not have any information on the dates where maintenance occurred. Patterns shift in the input vibration data alone account for all learning. Although this validates the innate use of adaptability in the model, this also reveals limitations. Directly presenting maintenance as a categorical/proportional input may help the model to differentiate enhanced fault recovery and superficial signal diversity. Maintenance data would allow for a different output, in such, learning can be done without expert interpretation. This would reduce possibility of misinterpretation when analyzing performance restoration.

3.8 Interpreting Model Accuracy and Generalization

Both mean squared error (MSE) and mean absolute error (MAE) from the test sets were employed to measure how accurate the model predicts as shown in Fig. 23. In each case, the pump datasets with the lowest errors—overall and per day—were Pump B with a MSE of 9 and MAE of 3.29 and Pump A with a MSE of 330 and MAE of 19. Meanwhile, datasets with faults or issues from pumps had much higher errors, with MSE between 469 and 1463 and MAE between 22.84 and 40.33 days. This finding matches what is expected because the model performs best when aging slowly and well when there is little background noise, but not well in rough and sudden environments.

The training was much more efficient after Bayesian regularization had been integrated into it. During all experimentations, regardless of quality of the signal data or length of the sets of data, training was steady and dependable. By controlling the overfitting during conditions in which data was in low availability, the algorithm enabled the network to genuinely generalize on unseen data. Stable convergence as well rationally set number of active weights showed the usefulness of the algorithm. For example, in works with limited amount of data, such as Pump F, the applied model utilized a smaller number of active weights, thus showing its adaptive capabilities to the complexity of the quality of the input.

In Fig. 24, it has shown how many effective parameters (weights + biases) are required by the trained model for each dataset. With fewer parameters, a network is

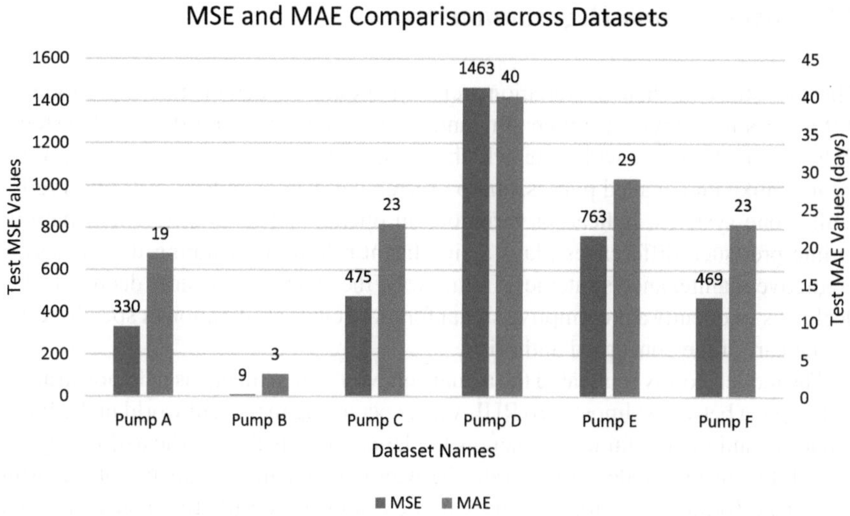

Fig. 23 Rate of MSE and MAE between all pumps

generally simpler which means it is less likely to get stuck with overfitting and better able to generalize with limited amount of data.

During the training process, the weights and biases in the network neurons are continually being changed. The use of error backpropagation and Bayesian regularization guides development of the model toward a simpler solution and away from excessively complex solutions that do not generalize well.

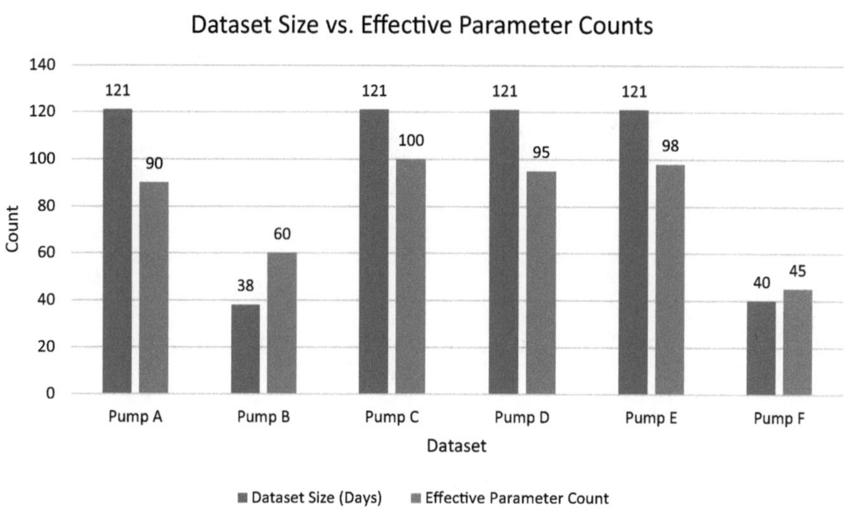

Fig. 24 Effective parameter and dataset size of all pumps

3.9 Insights from Degradation Trends

Through study of their degradation trends, the datasets demonstrated pronounced differences in behavior from healthy and faulty pumps. For non-damaged systems, RUL declined linearly with time, which indicated conditional operation and normal wear. Unlike undamaged pumps, sharp or irregular decline in RUL was observed in fault-prone systems, indicating variations in operational reliability. The capability to interpret such differences plays a significant role in determining the success of predictive maintenance strategies. In this way, the change from slow decline to fast decline is frequently a pre-emptive signal for more close observation or speedy action on the part of the concerned authority.

The model closely replicated the actual behavior of machinery as it demonstrated a near replica between almost zero RUL values occurring in the real world in the Pump E dataset and system failures; whereas the latter were distinct, indicated in Fig. 25. The ability of the model to illustrate the way maintenance is capable of restoring system performance is notable and captures both the destructive and restorative processes that are happening.

Also, significant improvements in RUL were identifiable (post-major maintenance) with reduced incidence when the equipment is operating in a normal environment. Unless the underlying problem is addressed, maintenance efforts had no substantial effect in improving RUL. On the contrary, if equipment reliability falls very quickly while maintenance becomes a continuing necessity, like in Pump D, replacing the equipment could be the best option instead of continuing the endless maintenance.

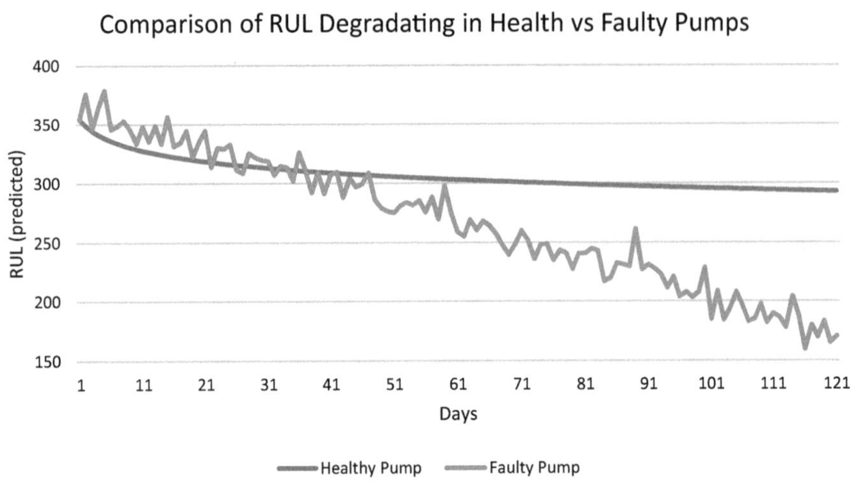

Fig. 25 Comparison of RUL in healthy pump versus faulty pump

3.10 Implications for Industrial Deployment

The predictive nature of the model shall have practical effects on actual-world industrial maintenance systems. The possibility of anticipating failures provides companies with a lever to launch preventive steps and thereby decrease the risk of catastrophic breakdowns and unplanned outage. Besides, the RUL predictions' interpretability allows the reliability engineers to check if the maintenance procedures recovered process equipment to the performance level expected of them.

Other than detection and response, the findings focus on that of constant monitoring and context-awareness. AI models work best if presented with high-quality high-frequency data backed by a maintenance log and domain knowledge. Incorporating such new light in the model construct would also take the robustness of the model to further heights especially in identifying the real failure and that of transition signal disturbances. The predictability of the model can over time become less dependent on expert analysis for standard situations, but expert consultation may still be necessary for complex or ambiguities situations through continuous improvement of gained operational experience and feedback.

Because the model relies entirely on data, it does not have a basic understanding of the physical rules of the pump behavior. Therefore, the model may disregard or read physical changes in the design of the pump or quality of the fluid wrongly, which is not trainable.

3.11 Insights from Literature Review

Much of what we have found here agrees with what has already been reported by others. As seen by Bruinsma et al. [6], under faulty conditions, vibration patterns deviate from those of healthy pumps just as the range and daily averages did for pumps D and E. The results of our study also fit with findings from Rajaoarisoa et al. [8], who revealed that explainable AI can identify and capture immediate repair gains. The excellent prediction achieved by the network, shown by a test MSE below 300, proves that Bayesian regularization should be considered, according to earlier studies. In addition, the differences in how well each pump was predicted are consistent with Irfan et al. [10], who indicated that fault intensity and sensor reliability impact the prediction's accuracy.

4 Conclusion

4.1 Summary of Research Purpose

Centrifugal pumps are crucial assets in many industrial facilities. This chapter explored how artificial intelligence could improve maintenance practices used for centrifugal pumps. The conventional maintenance policy, especially the reactive type is expensive and unreliable when failures caused by unexpected occurrences took place. The findings within the chapter show that predictive maintenance systems can be generated through machine learning. The machine learning uses vibration data from a real-world Omani oil and gas company, the machine used this data, to predict the residual useful life of pump and avoid pump failures that forces pump shutdown.

4.2 Literature Review Insights

The literature review unveiled the significance and shortcomings of the prevailing monitoring technique. Strategies like acoustic emission, motor current signal analyses, inductive wear detection, and IoT-enabled sensors give only the bits and pieces of malfunctions but are not good in bringing together the overall capabilities of producing real-time predictive abilities. However, as the review indicated, the evolving disciplines of AI and sensor technology have enabled it to become possible to construct more complex highly integrated condition monitoring systems. These gaps and trends provided a rationale for the focus of this study on neural network-based predictive maintenance.

4.3 Methodological Accomplishments

The methodology chapter reported the systematic procedures involved in developing a predictive model using information obtained from six pumps of different health conditions from their vibration data. The data were standardized to obtain primary indicators of mechanical degradation based on average and range metrics based on a daily interval. A historical window of 7 days is chosen to describe meaningful temporal patterns in the behavior of vibration. To train the model, a feedforward neural network with ten hidden neurons, regularized with Bayesian a generalization and stability strategy, was used. A new "DaysToFailure" metric was specified which combined information about vibration deviation and elapsed time and could approximate the useful remaining life without the need for actual failure times.

4.4 Performance and Interpretability of the Model

According to the findings, the results reported that the neural network performed well on datasets of healthy and faulty pumps. For healthy pumps, the model assumed a gradual degrading trend with exceptionally low mean squared error, clearly indicating particularly good tracking even from minor wear. More variability was observed in poorly working pumps, but with the overall reduction in useful life its general trend was also correctly recorded. Perhaps most importantly, the model demonstrated the capabilities to reflect the implications of maintenance interventions, which were not registered and known to the network. The rebound in remaining life shown for a pump when it was repaired even though the model had no direct knowledge of the maintenance. This indeed demonstrated the model's ability to learn from delicate changes in data and recognize system recovery.

4.5 Industrial Implications

The model illustrates a successful method of predictive maintenance that is useful for industrial purposes. Apart from alarms set up by thresholds, the system provides a forward-looking evaluation of degradation rates. This enables the engineers to make interventions prior to failure and displays if maintenance efforts achieved the desired effects. The way this approach makes the model capable of narrowing the gap between diagnostics and action-oriented decision-making as it considers the operational usage of the machinery as well as its natural deterioration with time.

4.6 Limitations and Future Outlook

The current study does, however, have limitations despite exciting results. It was fully dependent on data acquired using uniaxial vibration at the non-drive end of the pump and constrained its ability to determine types of faults. The dataset was, however, small and generalization of findings to other pump types or operating conditions may be restricted. In addition, although the model depended on the time-domain properties, adding frequency-area considerations through either fast Fourier transforms or wavelet approaches could find many fault signatures. Future research will require multi-sensor data fusion, real-time deployment through SCADA or IoT systems, and use of explainable AI on the model's decisions to show the maintenance team clear overviews of the same.

We should mention that among all these methods; acoustic emission is still seen as a valuable and versatile measuring device. Since this model is compatible with frequency mapping methods such as FFT, it could spot faults in the system early on. Enhancing learning machines with defect pattern features is likely to enable earlier discovery of possible defects in future versions.

4.7 Final Remarks

This chapter demonstrated that implementation and use of predictive maintenance based on artificial intelligence is not just possible, but also beneficial for monitoring centrifugal pumps. Through learning from previous performance of vibration, the developed model can predict trends in deterioration, help to plan maintenance, and minimize unplanned downtime. The research accelerates the development of the intelligent maintenance, having a significant bearing on scalable implementation in industrial sectors. Even though the model is not based on physical equations describing or mechanisms of a pump behavior, it utilizes statistical trends observed in the data to predict the time to remaining failure with a certain degree of precision.

Appendices

Appendix 1: DataPrep.m—Data Preprocessing Code

```
%% 1. Load Data
data = readtable('Pump Data.xlsx', 'Sheet', 'Pump A'); % Load vibration data from
Excel sheet
data.Date = datetime(data.Date);  % Convert date column to datetime format

%% 2. Compute Daily Statistics
uniqueDates = unique(data.Date);  % Identify all unique dates in the dataset
numDays     = numel(uniqueDates); % Count number of unique days
dailyStats  = table('Size',[numDays,3], ... % Initialize an empty table with 3
columns
                    'VariableTypes',{'datetime','double','double'}, ... % Specify
variable types
                    'VariableNames',{'Date','Average','Range'}); % Assign column
names
dailyStats.Date = uniqueDates;  % Assign unique dates to the 'Date' column

for i = 1:numDays
    idx = data.Date == uniqueDates(i);  % Logical index for rows matching the
current date
    vals = data.Y_AxisValue(idx);  % Extract Y-axis values for current date
    dailyStats.Average(i) = mean(vals); % Calculate and store daily average
    dailyStats.Range(i)   = range(vals); % Calculate and store daily range
end

%% 3. Generate Target Variable: DaysToFailure
baseline        = dailyStats.Average(1);  % Set baseline using first day's
average
k               = 0.5;  % Define sensitivity factor for deviation penalty
daysToFailure   = zeros(numDays,1); % Initialize array to hold Days to Failure
values

for i = 1:numDays
    elapsed          = days(dailyStats.Date(i) - dailyStats.Date(1));  % Days
passed since first reading
    deviation        = abs(dailyStats.Average(i) - baseline);  % Absolute
deviation from baseline
    daysToFailure(i) = max(0, 365 - elapsed - k * deviation); % Calculate Days to
Failure with penalty
end

dailyStats.DaysToFailure = daysToFailure;  % Add DaysToFailure column to the stats
table
```

Appendix 2: BuildFeatures.m—Feature Matrix Construction

```
%% 4. Build Feature Matrix X (14×N) and Target Y (1×N)
X = [];   % Initialize empty feature matrix
Y = [];   % Initialize empty target vector
for i = 8:numDays % Start from day 8 to ensure 7 previous days are available
    fv = [];  % Initialize empty feature vector for current day
    for j = i-7:i-1 % Loop over the previous 7 days
        fv = [fv, dailyStats.Average(j), dailyStats.Range(j)];  % Append average
and range as features
    end
    X = [X; fv]; % Add feature vector as a new row in X
    Y = [Y; dailyStats.DaysToFailure(i)]; % Append corresponding DaysToFailure
value to target vector
end
X = X';   % Transpose to get 14 × N (14 features, N samples)
Y = Y';   % Transpose to get 1  × N

% — Manual split fractions: 80% train, 20% validation —
numSamples = size(X,2);  % Total number of samples (columns in X)
nTrain     = floor(0.80 * numSamples); % 80% of the data for training
nVal       = floor(0.20 * numSamples); % 20% of the data for validation
trainInd   = 1:nTrain; % Indices for training data
valInd     = nTrain+1 : nTrain+nVal;  % Indices for validation data
testInd    = nTrain+nVal+1 : numSamples;  % Any remaining samples go to test set
(often empty)

% — Extract test set separately —
XT = X(:, testInd); % Feature matrix for test set
YT = Y(:, testInd);  % Target values for test set
```

Appendix 3: TrainEvaluate.m—Neural Network Training

```
%% 5. Create & Train Feedforward Network
hiddenLayerSize = 10; % Number of neurons in the hidden layer
net = feedforwardnet(hiddenLayerSize); % Create a feedforward neural network

net.input.processFcns= {'mapstd'}; % Normalize input features (zero-mean, unit
variance)
net.output.processFcns = {'mapstd'};   % Normalize target outputs similarly

net.divideFcn            = 'divideind'; % Use index-based data division

net.trainFcn = 'trainbr';  % Use Bayesian regularization for training
net.performParam.regularization = 0.1;   % Regularization factor to prevent
overfitting
net.divideParam.trainInd = trainInd; % Set training indices
net.divideParam.valInd   = valInd; % Set validation indices
net.divideParam.testInd  = testInd;  % Set test indices


[net, tr] = train(net, X, Y); % Train the network with input X and target Y
view(net);  % Display the network architecture

%% 6.a Evaluate on Test Set Only

YPredTest = net(XT); % Predict test set output
mseTest   = perform(net, YT, YPredTest); % Calculate mean squared error on test
set
fprintf('Test-set MSE: %.4f\n', mseTest); % Display the test-set MSE

% Calculate Mean Absolute Error (MAE) on test set
maeTest = mean(abs(YT - YPredTest));
fprintf('Test-set MAE: %.4f days\n', maeTest);


%% 6b. Display Predictions from Day 8 to Day numDays
% Run the network on all feature samples
YPredAll = net(X);     % Predict DaysToFailure for all samples
YPredAll(YPredAll < 0) = 0; %Ensure values do not go below 0

fprintf('\nPredictions for Days 8 → %d:\n', numDays);
for i = 1:size(YPredAll,2)
    dayIdx = i + 7;    % Convert index back to original day numbers
    fprintf('Day %3d → %.1f days to failure\n', dayIdx, YPredAll(i)); % Print
prediction
end

%% 7. Single-Sample Prediction for the Last Day
newSample = X(:, end); % Select features for the last available day
predLast  = net(newSample); % Predict days to failure for that last day
fprintf('\nOn day %d → %.1f days to failure\n', numDays, predLast); % Output the
result
```

References

1. Alfayez L, Mba D (2005) Detection of incipient cavitation and determination of the best efficiency point for centrifugal pumps using acoustic emission. Proceed Instit Mech Eng E J Proc Mech Eng 219(4):327–344. https://doi.org/10.1243/095440805x33225

2. Neill GD, Reuben RL, Sandford PM, Brown ER, Steel JA (1997) Detection of incipient cavitation in pumps using acoustic emission. Proc Inst Mech Eng Part E: J Process Mech Eng 211(4):267–277. https://doi.org/10.1243/0954408971529737
3. Sun H, Zhang W, Li Y, He X, Feng Y (2024) Cavitation detection via motor current signal analysis for a centrifugal pump in the pumped storage pump station. J Energy Stor 81:110417. https://doi.org/10.1016/j.est.2024.110417
4. McKree KK (n.d.) A review of major centrifugal pump failure modes with application to the water supply and sewerage industries. Curtin University
5. Hua M, Liu S, Zheng Q, Huang B, Zhang X (2023) Failure analysis and structural fatigue resistance design of multistage centrifugal pump shaft. Eng Fail Anal 153:107545. https://doi.org/10.1016/j.engfailanal.2023.107545
6. Bruinsma S, Wang D, Meijer L (2024) Motor current and vibration monitoring dataset for various faults in an E-motor-driven centrifugal pump. SSRN. https://doi.org/10.2139/ssrn.4617449
7. Mohammed A (2023) Data-driven model for predicting pump failures in the oil and gas industry. Eng Fail Anal 145:107019. https://doi.org/10.1016/j.engfailanal.2022.107019
8. Rajaoarisoa L, Zhang C, Wang P, Tsang E (2025) Decision-making systems improvement based on explainable artificial intelligence approaches for predictive maintenance. Eng Appl Artif Intell 139:109601. https://doi.org/10.1016/j.engappai.2024.109601
9. Lu S, Chen Y, Liu Z, Li X (2023) Edge computing on IoT for machine signal processing and fault diagnosis: a review. IEEE Internet Things J 10(13):11093–11116. https://doi.org/10.1109/JIOT.2023.3239944
10. Irfan M, Hussain S, Khalid S, Tariq M (2021) A novel feature extraction and fault detection technique for the intelligent fault identification of water pump bearings. Sensors 21(12):4225. https://doi.org/10.3390/s21124225
11. Şevik S (2022) Impact of maintenance on pump performance [conference paper]. In: Proceedings of the 6th international conference on innovative studies of contemporary sciences
12. Yıldız GB, Soylu B (2023) Integrating preventive and predictive maintenance policies with system dynamics: a decision table approach. Adv Eng Inform 56:101952. https://doi.org/10.1016/j.aei.2023.101952
13. Chen L, Zhang Y, Wang H, Xu T (2022) Monitoring and predictive maintenance of centrifugal pumps based on smart sensors. Sensors 22(6):2106. https://doi.org/10.3390/s22062106
14. Mohtar N (2013) Vibration analysis of centrifugal pump [Bachelor's dissertation, Unknown University]
15. Menanno M, Salsano A (2023) Centrifugal pump fault diagnosis using a predictive maintenance model. In: Proceedings of the international conference on industrial engineering and operations management. https://doi.org/10.46254/eu6.20230422
16. Ucar A, Karakose M, Kırımça N (2024) Artificial intelligence for predictive maintenance applications: key components, trustworthiness, and future trends. Appl Sci 14(2):898. https://doi.org/10.3390/app14020898

Parameter Identification of Empirical Models for Head Estimation of Electrical Submersible Pumps for Viscous Fluids

Hadab Hadabi, Morteza Mohammadzaheri, and Ali Al-Humairi

Abstract ESPs are important in oil production, but their efficiency drops a lot when they handle viscous fluids. When these conditions exist, the usual water-based performance curves are not reliable and can result in wrong predictions of pump head and flow rate. This chapter demonstrates how to identify and optimize the parameters of KSB (Kreiselpumpen-Lexiko, KSB, 1989) and Gülich (Centrifugal pumps, 2008) models to enhance the accuracy of predicting ESP performance in viscous conditions. Experimental data from a TE2700 ESP working with viscosities up to 520 cP and speeds from 1800 to 3500 rpm were used to calculate and optimize the correction factors for head and flow using Nelder–Mead and Levenberg–Marquardt algorithms. The optimized KSB model performed well, reducing errors significantly and achieving R^2 values above 0.98 and MAPE below 13%. The attempt to linearize nonlinear models was not successful because the models were too complex. In addition, a multi-linear regression model was built which is simple but does not generalize well. The study introduces a detailed computational system to enhance ESP modeling in viscous conditions and suggests future research topics such as multiphase flow modeling and AI-based regression.

Keywords Electrical submersible pumps · Viscosity correction · Empirical models · Parameter optimization · Nonlinear modeling · Nelder–Mead · Levenberg–Marquardt · KSB · Gülich · Linear regression

Abbreviations

ESP	Electrical submersible pump
BEP	Best efficiency point
MAE	Mean absolute error
RMSE	Root mean square error
MAPE	Mean absolute percentage error

H. Hadabi · M. Mohammadzaheri (✉) · A. Al-Humairi
German University of Technology in Oman, Halban, Sultanate of Oman
e-mail: morteza.zaheri@gutech.edu.om

A. Azizi (ed.), *Smart Diagnostics and Predictive Maintenance*, Emerging Trends in Mechatronics, https://doi.org/10.1007/978-981-95-3921-5_6

ARE	Absolute relative error
R^2	Coefficient of determination
CFD	Computational fluid dynamics
PBM	Population balance model
MLR	Multiple linear regression
LM	Levenberg–Marquardt
NM	Nelder–Mead
VSD	Variable speed drive
RPM	Revolutions per minute
TUALP	The university of Alberta lift project
S.G	Specific gravity
FNN	Fully connected neural network
SVR	Support vector regression
GPR	Gaussian process regression
RBF	Radial basis function
GBM	Gradient boosting machine
KRR	Kernel ridge regression
GAM	Generalized additive model

Symbols

H	Head (m or ft)
Q	Flow rate (m^3/s or BPD)
η	Efficiency
P	Power (W)
N	Rotational speed (rpm)
ω	Rotational speed (rad/s or rpm)
ω_s	Specific speed
$\omega_{s,\mathrm{op}}$	Operational specific speed
ν	Kinematic viscosity (m^2/s or cSt)
μ	Dynamic viscosity (Pa s or cP)
ρ	Fluid density (kg/m^3)
g	Gravitational acceleration (m/s^2)
u_1, u_2	Inlet and outlet velocities (m/s)
p_1, p_2	Inlet and outlet pressures (Pa or N/m^2)
u_1, u_2	Elevation (m)
r_2	Outer radius of impeller (m or mm)
C_Q	Flow correction factor
C_H	Head correction factor
C_η	Efficiency correction factor
ξ	Off-BEP correction factor
B_{HI}	Viscosity-dependent parameter in KSB model
B	Dimensionless correction parameter

$Re_{evdocia}$	Modified Reynolds number (Stepanoff model)
Re_{ω}	Rotational Reynolds number
$Re_{Gülich}$	Gülich Reynolds number
Re_{Ofuchi}	Ofuchi Reynolds number
θ	Parameter vector (used in optimization)
J	Jacobian matrix
λ	Damping factor (Levenberg–Marquardt)
R	Residual vector
X	Exponent in correction factor equations
Z	Exponent for flow correction in
β_n	Regression coefficients (in MLR model)
E	Error term (in regression)
x_{norm}	Normalized value
M	Mean (in normalization)
Σ	Standard deviation (in normalization)

1 Introduction

1.1 Background and Context

Electrical submersible pumps (ESPs) are important in today's oil industry because they help extract oil from difficult reservoirs in a reliable way [1]. They can lift fluids to great heights due to multistage centrifugal systems that utilize downhole electric motors to convert electrical energy into mechanical and hydraulic pressure [1]. Armais Arutunoff's invention in 1926 marked the beginning of this technology, which revolutionized oil extraction [2].

Since that time, ESP technology has advanced with many innovations [3]. The standard components are still the electric motor, protector, pump, and cable [2]. In the 1950s, mechanical seals were introduced to protect motors, and in the 1970s, rotary gas separators and variable speed drives (VSDs) were developed to improve performance and flexibility under high gas content [2].

Because ESPs can handle a wide range of operations, from 200 to 60,000 barrels per day and up to 15,000 ft, they are used in many industries [4]. ESPs are more flexible and efficient than sucker rod pumps and gas lifts and they can be used in both onshore and offshore settings [5].

In labs, the main performance factors of a pump, such as pump head, flow rate, and efficiency, are tested using water, which helps create standardized curves needed for field use [2]. Pump head is especially important for choosing the right stage and motor for the reservoir [6].

Still, the behavior of ESP is influenced by viscosity. When the viscosity of the fluid increases, it causes more friction inside the pump and reduces its performance [7]. In

1946, Ippen introduced correction factors for water-based predictions. Later studies by Trevisan and Prado found that viscosity greatly reduces the pressure generated.

For this reason, empirical modeling is now essential for predicting how ESPs will work with viscous or multiphase fluids in real-world situations. Gülich [6] and Stepanoff [8] improved these models, which helped close the gap between lab and field results and made predictions more accurate.

1.2 Problem Statement

The production of wells with declining reservoir pressure depends on effective artificial lift systems, making electrical submersible pumps (ESPs) increasingly preferred due to their high flow capacity and deep-set capabilities (Trevisan and Prado 2011). Standard ESP performance curves are developed using water during laboratory testing, such as the typical manufacturer curve shown in Fig. 1. Still, these water-based curves do not accurately reflect the actions of viscous and multiphase fluids that are common in oilfield operations [3]. It was pointed out that these testing methods can lead to major errors in predicting oil production, mainly in heavy oil or complicated reservoirs.

The main issue is that viscosity changes the way the pump's internal flow works which results in changes in head, efficiency, and flow rate [9]. Unsteady transitions and gas entrainment negatively affect the reliability of the empirical models. In addition, correction factor-based empirical models have limitations because of their basic assumptions. Moreover, linearized models are not able to handle the nonlinear effects that come from changes in fluid properties. However, nonlinear models, however,

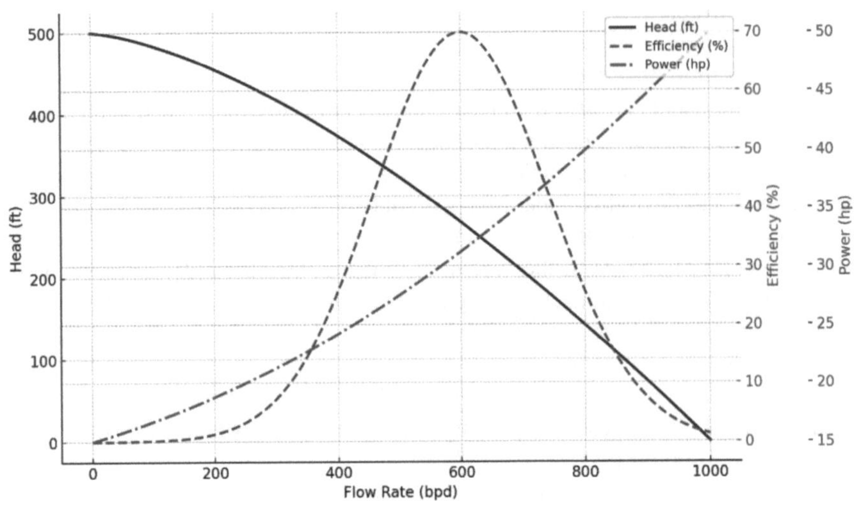

Fig. 1 Example of a typical performance curve of an electrical submersible pump (ESP)

make it is difficult to estimate parameters of nonlinear models [10]. There is still a lack of general models that can accurately predict ESP performance in complex flow situations, which is why this research is needed.

1.3 Objectives

The rising complexity of oil extraction involving high-viscosity fluids and multiphase flows demands precise ESP performance prediction [9]. Traditional empirical models built on volute centrifugal pump datasets are widely used but fall short in predicting the performance of multistage ESP systems under demanding conditions [9, 11]. These limitations in head, flow rate, and efficiency prediction result in improper system design and higher operational costs [11].

This research establishes a structured computational approach to evaluate and optimize empirical models, including [8, 11–16]. Despite their frequent citation, these models lack thorough assessment and linear transformation for robust parameter estimation. The primary objective was to attempt linearization of these models to enable straightforward estimation, as it was proved infeasible, nonlinear optimization is to be performed. As a contribution, a custom linear model tailored to the data is also be developed.

Using MATLAB, these models are implemented with experimental data from Shi et al. [9] across various viscosities and operating speeds. The study excludes efficiency predictions and focuses on head and flow rate behavior. Parameter identification relies on minimizing error indicators such as MAPE, MAE, RMSE, and R^2 [17]. Validation is done to compare predicted head values to experimental benchmarks.

Recent findings show that standard correction correlations yield unreliable field results [9], necessitating ESP-specific recalibration [11]. The outcome is a refined modeling strategy for improved pump selection, system efficiency, and simulation integration under viscous flow conditions.

1.4 Scope and Limitations

This study focuses on parameter identification and linear model development for empirical head estimation models applied to electrical submersible pumps (ESPs) under viscous fluid conditions. Only empirical models are considered, excluding analytical, mechanistic, and hybrid types. Several models including the ones introduced in [8, 9, 11, 13, 14, 16], were investigated due to their specific relevance to viscous head estimation. The work centers on pump specific model optimization using experimental data for calibration. MATLAB is the main platform for implementation and analysis.

The experimental data originate from a 14-stage TE2700 ESP operating with viscosities between 1 and 520 cP, flow rates between 15 and 50 m^3/day, and speeds from 1800 to 3500 rpm. This pump, equipped with radial impellers and diffusers, serves as a representative ESP model [9]. However, the findings are limited to this configuration and cannot be generalized to other ESP designs without further validation. The dataset used was collected before this research [9], limiting the scope to its predefined operational conditions. Only single-phase fluid data are used, multiphase flow, gas–liquid interaction, and solids are excluded but acknowledged as important for future work.

This study intentionally excludes nonlinear fitting techniques and CFD-based modeling to focus on accessible, computationally efficient linearization strategies. The scope is limited to head and flow rate prediction, with efficiency analysis outside the current research boundaries. Additionally, no hardware testing or field validation was conducted, which will be necessary for full practical verification.

1.5 Chapter Structure

The chapter follows a structured sequence of four sections, progressing from theoretical foundations to methodology, analytical results, and conclusions.

Section 1 introduces electrical submersible pumps (ESPs), emphasizing their significance in artificial lift and the challenge of predicting head performance under viscous conditions. The section outlines the research problem, objectives, scope, limitations and the organization of the chapter. It explores prior research on ESP operation in high-viscosity environments and reviews the development of empirical modeling techniques. It discusses the limitations of current correction factor methods, the absence of generalizable linear models, and the challenges of parameter identification. The section highlights the motivation behind linearizing nonlinear models and frames the research gaps addressed in this study.

Section 2 presents the methodology, describing the selection and linearization of six empirical models, implementation in MATLAB, and parameter optimization using Nelder–Mead and Levenberg–Marquardt methods. It details the use of experimental data from a TE2700 ESP, outlines the data preprocessing approach, and defines the error metrics used for evaluation.

Section 3 analyses the results of the parameter identification, including visual and statistical comparison of predicted and experimental head values. It presents both training–testing and full-dataset optimization results, model behavior across varying viscosities and rpms, and discusses the attempt to apply a data-specific multiple linear regression (MLP) model as additional analysis.

Section 4 concludes the chapter by summarizing key findings and reflecting on the limitations encountered, particularly the infeasibility of linearizing certain models. It discusses the added contribution of nonlinear optimization for improving prediction accuracy and suggests future directions, including expanding datasets, exploring multiphase effects, and developing more adaptable empirical models. In the end, section contains the references.

Literature Review

1.6 *Viscosity Effects on Pump Performance*

1.6.1 Impact of Viscosity on Head and Flow Rate

Various studies on numerical and experimental grounds confirm that significant degradation in the head and flow rate performance of ESP is due to increasing fluid viscosity. Zhu et al. [18] reported that the boosting pressure of a multistage ESP decreases by approximately 30–40% when oil viscosity increases from 10 to 100 cP, and becomes almost ineffective beyond 200 cP, based on CFD simulations and experimental measurements conducted on a seven-stage ESP system. These findings are verified by Sirino et al. [19] who stated that pressure head curves shift downward with viscosity increases, and these shifts are accompanied by reduced efficiency and distorted flow structures. While Zhu et al. [18] focused on the quantitative head loss, on the other hand, Sirino et al. [19] emphasized that internal causes of head loss are due to separation and recirculation zones.

This trend using a three-stage ESP was validated by Ofuchi et al. [20, 21], where higher viscosities produced lower flow rates for a given head, highlighting a near-linear behavior in the head to flowrate curves at extreme viscosities. Thus, the shift from turbulent to laminar flow regimes is the resulting behavior. Furthermore, Shojaeefard et al. [22] stated that the head generation becomes progressively ineffective due to internal flow obstruction and momentum losses via an increase in viscosity. Nevertheless, each paper grants specific intuitions, their summaries are bound to particular pump rotational speeds and geometries. To point out, the current study compares findings of another dataset, presenting further analysis.

1.6.2 Viscous Influence on Efficiency and Energy Losses

Viscosity not only affects pressure head but also drastically alters the energy efficiency profile of ESPs. Zhu et al. [18] demonstrated that increasing oil viscosity leads to deterioration of the best efficiency point (BEP), shifting it toward lower flow rates and heads. This degradation results from intensified internal friction and the emergence of energy-dissipating flow structures. Supporting this, Ofuchi et al. [21] confirmed through CFD modeling that increased viscosity produces higher wall shear stress and turbulence intensity within the pump casing, thereby escalating mechanical and hydraulic losses.

Shojaeefard et al. [22] further showed that viscous effects increase disc friction and leakage flow through clearances, reducing hydraulic efficiency by up to 20% in extreme cases. These findings align with [23], who observed that flow behavior changes abruptly under viscous regimes, leading to nonlinear head responses such as the sudden-rising head phenomenon. Together, these results stress the inadequacy of standard design curves and performance charts that are based on water testing.

In real-world applications, these losses translate to increased operating costs, as ESPs must work harder to maintain target flow rates, often requiring higher power inputs. Ofuchi et al. [21] noted that greater torque demand on the motor increases thermal loading on bearings, which in turn shortens equipment life and raises maintenance requirements. These conclusions are also echoed by Shojaeefard et al. [22], who demonstrated that higher viscosity increases the thermal burden on rotating components, while [23] showed that abrupt shifts in flow regimes can introduce additional instability and stress. As an outcome, viscosity imposes equally operational inadequacies and business obligations on field operations, resulting in performance degradation.

1.6.3 Flow Structure Alterations and Recirculation Effects

Viscosity significantly distorts the internal flow structure of ESPs. Zhu et al. [18] and Stel et al. [24] both documented that recirculation zones being formed near the impeller trailing edges were revealed through their CFD simulations, contributing to localized head losses and impeller blade loading. Ofuchi et al. [20, 21] observed that flow misalignment entering downstream stages leads to cumulative degradation due to stage-to-stage interaction of flow separation. This is compounded by findings from Sirino et al. [19], who demonstrated that increasing viscosity intensifies wake formation and backflow, particularly in multistage configurations.

Stel et al. [24] expanded on this by showing that viscosity-induced velocity skewing increases radial thrust on the shaft, resulting in mechanical fatigue and alignment issues over extended operation. For instance, [22] validated these structural impacts by documenting secondary vortex formations and boundary layer thickening under high-viscosity circumstances. These vortices not only disrupt flow uniformity but also contribute to amplified vibration and noise, potentially accelerating component wear.

Pressure variations became more obvious under viscous flow, as they introduced instability into ESP operations through flow visualizations [22]. For instance, uncertainties degrade the consistency of standard sensors used for flow and pressure monitoring and make control systems less receptive. These observations are strengthened by Stel et al. [24], who stated that pressure profiles are sensitive to viscosity changes, and Ofuchi et al. [20], who highlighted increased flow pulsation due to density–viscosity coupling. Despite these valuable observations, existing design methodologies seldom account for such internal flow dynamics, highlighting a critical gap in conventional ESP design practices.

1.7 *Review of Existing Empirical Models for Head Estimation*

1.7.1 Overview of Performance Prediction Approaches for Electrical Submersible Pumps

Operation optimization under varied flow conditions is a critical aspect in ESPs' performance prediction [25]. For instance, several methods have stood out to forecast ESP efficiency, head, and flow in both ideal and non-ideal conditions. These approaches broadly fall into four categories: traditional empirical fitting models, mechanistic models based on physical principles, computational fluid dynamics (CFD)-based simulations, and machine learning-based models [26].

The earliest techniques utilized in ESP performance estimation include traditional empirical methods [11]. For example, these models have historically been widely adopted in industry practice, while they use regression-based calibration against experimental pump curves. A study by Monte Verde et al. [11] compares various empirical formulations like those of ANSI-HI [12], KSB [14], and Monte Verde [15], showing that although empirical models can MAPE values below 5% when recalibrated, their generalization outside the BEP is limited [11]. Furthermore, Ofuchi et al. [16] and Stepanoff [8] models lack flexibility in demonstrating nonlinear behaviors, yet they remain widely referenced [11].

The use of physical laws and dimensional analysis to estimate ESP performance under viscous flow regimes falls below analytical alternatives known as mechanistic models. For example, according to Zhu et al. [27] based on impeller geometry, operating speed, and energy balance equations dimensionless coefficients are defined as in the flow coefficient (ϕ), head coefficient (ψ), power coefficient (Π), and efficiency (η) obtained. These models excel at identifying trends in pump behavior using first principles, particularly in single-phase flow and moderate gas conditions [27]. While [28] states that mechanistic models degrade without extensive calibration data, but the utility of physically grounded models for their predictive consistency.

Computational fluid dynamics (CFD) techniques represent a powerful approach for simulating ESP operation under complex conditions. CFD coupled with population balance models (PBM) has been applied to model gas–liquid two-phase flows in centrifugal and ESP systems. Zhu et al. [29] demonstrate that traditional Eulerian–Eulerian models are outperformed by CFD-PBM coupling in describing dynamic bubble behavior and pressure increase in ESPs under high gas fractions. Their simulation pointed out that performance degradation occurs at a void fraction above 15%, as strong interaction between vapor bubbles and impeller channels was observed [29]. Although acquiring a highly accurate CFD model requires expensive computational power, sensitive mesh quality analysis, defined boundary conditions, and turbulence model selection, thus makes them more fit for design optimization rather than routine forecasting of field pump performance [27].

Machine learning models, particularly deep learning algorithms, have recently gained momentum due to their ability to model highly nonlinear interactions between operating parameters and ESP performance. Studies such as those by Mohammadzaheri et al. [30] and Zhu et al. [29] highlight how deep neural networks can dynamically adapt to changing field conditions and outperform conventional methods in predictive accuracy. Seven learning methods, generalized additive models (GAMs), support vector regression (SVR), Gaussian process regression (GPR), radial basis functions (RBF), gradient boosting machines (GBMs), Kernel ridge regression (KRR), and fully connected neural networks (FNNs) were evaluated in terms of their accuracy, adaptability, and computational efficiency [29]. To conclude, FNNs showed superior generality even with small training data. Conversely, complications arise with overfitting, limited interpretability, and architecture tuning remain [29].

Furthermore, Abdalla et al. [31] used a mix of support vector regression, gradient boosting, and convolutional neural networks to predict flow rate in ESP-lifted wells, demonstrating that hybrid models can provide robust real-time forecasting capabilities [31]. Nonetheless, these models often require a large volume of clean training data and may lack physical interpretability compared to analytical approaches.

1.7.2 Affinity Laws and Their Role in ESP Performance Scaling

Changes in flow rate, head, and power in pump engineering can be estimated using a widely utilized classical tool known as affinity laws, these estimations are a result of variations in pump speed or impeller size correlations [32]. These correlations are retrieved through the assumption of Newtonian, inviscid fluid flow and geometric similarity. As presented in Simpson and Marchi [32], the standard affinity laws are defined as:

$$\frac{Q_1}{Q_2} = \frac{N_1}{N_2},\ \frac{H_1}{H_2} = \left(\frac{N_1}{N_2}\right)^2,\ \frac{P_1}{P_2} = \left(\frac{N_1}{N_2}\right)^3, \tag{1}$$

where Q is the flow rate, H is the head, P is power, and N is the pump's rotational speed [32].

The affinity laws effectiveness can be seen in low-viscosity fluids such as water. Simpson and Marchi [32] state that affinity laws can predict head and flow precisely in systems that are water-based, particularly in variable speed pumping (VSP) applications. Nonetheless, unreliable efficiency and power predictions were observed when the pump rotational speed was significantly reduced [32]. For example, in a small 5.5 kW centrifugal pump tested at 2000 rpm, the predicted BEP dropped from 56 to 52%, highlighting this discrepancy [32].

Increase in fluid viscosity significantly deteriorates affinity laws validity. For example, [7] presented experimental tests done on radial and semi-axial ESPs working with glycerin as a fluid at viscosities between 60 and 1020 cP revealing noticeable deviations. At 60–720 cP, the deviation in flow rate from HI-USA chart predictions [33] exceeded 20% at the BEP, while at 1020 cP, efficiency deviation exceeded 35% at the same operating point [7]. These results confirm that the affinity laws are no longer valid above 100 cP, especially when applied without correction factors [7].

Experts have studied several ways to improve the limitations of affinity laws. For example, [34] suggested including Reynolds number effects by adding an exponent that represents the viscosity effects. Their modified law allowed them to predict fluids from 1 to 400 cP and rotational speeds from 2400 to 4200 rpm which improved the accuracy of their head and efficiency predictions [34]. Furthermore, [32] related an empirical correction developed by Sarbu [35] to enhance the efficiency and prediction accuracy in VSP systems. They were able to reduce through their method the BEP divergence down to <2.5% from over 7.6%. Therefore, such empirical correction models are best fit for use in small pump purposes and hydraulic solvers under variable speed conditions [32].

In summary, the classical affinity laws offer a reliable framework for centrifugal pump performance prediction under ideal, low-viscosity fluid conditions. Though, beyond 100 cP the predictive accuracy declines sharply, as evidenced by Amaral et al. [7] data. Affinity laws' general applicability across all ESP flow regimes and geometries remains limited, even though modified empirical corrections and scaling formulations show promising results. Direct reliance on affinity laws without correction may result in substantial errors in head, flow, and efficiency estimations for high-viscosity fluids.

1.7.3 Empirical Correction Factors and Common Formulae

When centrifugal pumps such as electrical submersible pumps (ESPs) operate under viscous fluid conditions, their performance deviates substantially from standard water-based predictions [36]. To address this, empirical correction factors are employed to adjust water-based pump curves, enabling performance prediction under viscous operating conditions. These correction factors are defined as ratios of viscous fluid performance to that of water under the same conditions [37]:

$$C_Q = \frac{Q_{\text{vis}}}{Q_w}, \tag{2}$$

$$C_H = \frac{H_{\text{vis}}}{H_w}, \tag{3}$$

$$C_\eta = \frac{\eta_{\text{vis}}}{\eta_w}, \tag{4}$$

where Q, H, and η represent flow rate, head, and efficiency respectively, with subscripts "vis" and "w" denoting viscous and water conditions. These factors are central to several empirical models, allowing users to estimate the degraded performance of pumps handling non-Newtonian or high-viscosity fluids without conducting extensive physical testing [37].

Among the early and widely referenced models is the [8] formulation, which uses a modified Reynolds number (Eq. 6) to calculate correction factors based on specific speed (ω_s) and viscosity. This model assumes the pump head decreases as the viscosity increases at a constant speed. Thus, specifically at BEP, the specific speed (ω_s) remains constant. As described by Monte Verde et al. [11], the modified Reynolds number is defined as:

$$\mathrm{Re}_{\mathrm{evdocia}} = 6.0345 \frac{\omega Q_{w,\mathrm{BEP}}}{\nu \sqrt{H_{w,\mathrm{BEP}}}}, \tag{5}$$

where ω is the rotational speed in rpm, ν is the kinematic viscosity in cSt, $H_{w,\mathrm{BEP}}$ is the head of the pump at BEP in ft, and $Q_{w,\mathrm{BEP}}$ is the water flow rate at BEP in bpd. While the specific speed ω_s equation is given by:

$$\omega_s = \frac{\omega\sqrt{Q_{w,\mathrm{BEP}}}}{\left(\mathrm{gH}_{w,\mathrm{BEP}}\right)^{3/4}} = \frac{\omega\sqrt{Q_{\mathrm{vis,BEP}}}}{\left(\mathrm{gH}_{\mathrm{vis,BEP}}\right)^{3/4}}, \tag{6}$$

where g is gravitational acceleration constant in m/s^2. This non-dimensional parameter ($\mathrm{Re}_{\mathrm{evdocia}}$) serves as the basis for computing head correction at BEP ($C_{H,\mathrm{BEP}}$), and we can see in the below Eq. (8) b and c the empirical dimensional parameters, as shown in Table 1:

$$C_{H,\mathrm{BEP}} = 1 - e^{\left(-b\mathrm{Re}_{\mathrm{evdocia}}^{c}\right)}. \tag{7}$$

The author then generalized Eq. (7) to get the equation for $C_{Q,\mathrm{BEP}}$ which is a function of $C_{H,\mathrm{BEP}}$ and a an empirical constant parameter, as shown in Table 1, defined as:

$$C_{Q,\mathrm{BEP}} = C_{H,\mathrm{BEP}}^{a}, \tag{8}$$

Another major contribution is the [14] model, which introduces the viscosity-dependent parameter B_{HI} to characterize the combined effects of flow, head, and fluid properties, which are calculated at water BEP:

$$B_{\mathrm{HI}} = \frac{480(\nu)^{0.5}}{\left(Q_{w,\mathrm{BEP}}\right)^{0.25}\left(gH_{w,\mathrm{BEP}}\right)^{0.125}}, \tag{9}$$

where the kinematic viscosity (ν) measured in m^2/s, water flow rate at BEP in m^3/s, water head at BEP in m, and g is the gravitational acceleration in m/s^2. Then the parameter B_{HI} is tuned to accompany the effect of the specific speed ω_s (Eq. 9), giving us the B parameter:

Table 1 Empirical constant values

Empirical model	Parameters	Original	Optimized
Stepanoff [8], TUALP	a	1.5	2.06
	b	0.033823	0.05484
	c	0.36769	0.30966
KSB [14]	a	15	10.41
	b	0.013	0.039
	c	0.165	0.113
	d	4	5.323
	e	0.25	0.390
	f	0.75	0.643
Gülich [13]	a	1.5	0.507
	b	6.7	12.09
	c	0.735	0.709
ANSI-HI [12]	a	0.5	0.429
	b	0.0625	−0.084
	c	0.375	0.243
	d	0.25	0.162
	e	2.71	2.652
	f	0.165	0.144
	g	3.15	3.829
Monte Verde [15]	a	1.5	0.462
	b	145.965	10.246
	c	1.139	0.741
	f	41.651	26.347
	g	0.688	0.663
Ofuchi et al. [16]	a	1.5	2.084
	b	4.462	5.306
	c	0.695	0.735

Monte Verde et al. [11]

$$B = B_{\mathrm{HI}}\sqrt{\frac{a}{52.933\omega_s}}, \tag{10}$$

using B, the flow correction factor is computed with a combined logarithmic and exponential formulation, resulting in:

$$C_Q = \left(\frac{a}{52.933\omega_s}\right)^{b\cdot B} e^{-c(\log_{10} B)^d}, \tag{11}$$

Here the *a*, *b*, *c*, and *d* are empirical parameters, as shown in Table 1. The head correction factor includes a scaling coefficient ξ that accounts for deviations from BEP:

$$\xi = 1 - 0.014(B_{\text{HI}} - 1).\left[\left(\frac{Q_w}{Q_{w,\text{BEP}}}\right) - 1\right], \tag{12}$$

$$C_H = (e + f \cdot C_Q)\xi, \tag{13}$$

where coefficients *e* and f are empirical constant as well, as shown in Table 1. The KSB [14] also provided a methodology to calculate the efficiency correction factor based on the specific speed and the *B* parameter, unlike the Stepanoff [8],which lacks that capability. Thus, the KSB [14] equations provide a robust way to compute corrections for both BEP and off-BEP conditions [11].

The Gülich [13] formulation takes a more physics-based approach, defining correction factors using rotational Reynolds numbers, as Karassik et al. [38] prediction method is based on semi-empirical energy loss and experimental findings [6]. The Gülich [13] dimensional Reynolds number $\text{Re}_{\text{Gülich}}$ is based on specific speed ω_s and the dimensional rotational Reynolds number Re_ω, given as:

$$\text{Re}_\omega = \frac{\omega r_2^2}{\nu}, \tag{14}$$

where r_2 is the outer diameter of the impeller. Then, the $\text{Re}_{\text{Gülich}}$ is defined as:

$$\text{Re}_{\text{Gülich}} = \text{Re}_\omega \omega_s^a, \tag{15}$$

here *a* is an empirical coefficient, as shown in Table 1. The BEP head correction factor is expressed as a function of $\text{Re}_{\text{Gülich}}$ and *x* parameter indicated below:

$$x = -\left(\frac{b}{\text{Re}_{\text{Gülich}}^c}\right), \tag{16}$$

$$C_{H,\text{BEP}} = \text{Re}_{\text{Gülich}}^x, \tag{17}$$

coefficients *b* and *c* are empirical constants, as shown in Table 1. Gülich [13] assumes that the head correction factor at BEP is equal to the flow rate correction factor, which applies to different flow rates and BEP.

$$C_Q = C_{H,\text{BEP}}, \tag{18}$$

For the head correction factor at different flow rates compared to that at BEP, the author proposed:

$$C_H = 1 - \left(1 - C_{H,\mathrm{BEP}}\right)\left(\frac{Q_w}{Q_{w,\mathrm{BEP}}}\right)^{0.75}, \tag{19}$$

As explained by Monte Verde et al. [11], the Gülich [13] allows estimation of efficiency correction factors across a wide range of flow rates, similar to the KSB [14] model.

The Hydraulic Institute provided an empirical model based on a vast dataset for different pump sizes operating under viscous fluid as the medium [12]. Monte Verde et al. [11] state that only the functional relations are omitted due to copyright reasons for the ANSI-HI [12] empirical model. All the correction factors are based on B the independent parameter, the correction factors are defined as:

$$B = F_1\left(Q_{w,\mathrm{BEP}}, C_{H,\mathrm{BEP}}, \omega, \nu, a, b, c, d\right), \tag{20}$$

$$C_{H,\mathrm{BEP}} = F_2(B, e, f, g), \tag{21}$$

$$C_H = F_3\left(C_{H,\mathrm{BEP}}, Q_{w,\mathrm{BEP}}, Q_w\right), \tag{22}$$

The above parameters a, b, …, g are empirical coefficients which are calculated constants, as shown in Table 1. The ANSI-HI [12] model runs under the same assumption made by Gülich [13] that the $C_{H,\mathrm{BEP}}$ is equal to C_Q outside and inside BEP flow ranges as shown in Eq. (19). The ANSI-HI [12] methodology also predicts the efficiency correction factor as the Gülich [13] and KSB [14] models.

Monte Verde [15] introduced further refinement using experimental ESP data, proposing viscosity-adjusted correction factors, as the empirical model is based on the Gülich [13] model, while varying relations were introduced for the flow rate, head, and efficiency. The head correction factor is based on Eqs. (17) and (18), as the flow rate correction factor is based on the z parameter, given as:

$$z = -\left(\frac{f}{\mathrm{Re}_{\mathrm{Gülich}}^{g}}\right), \tag{23}$$

where f and g are empirical constants, as shown in Table 1, the flow correction factor at BEP is:

$$C_{Q,\mathrm{BEP}} = \mathrm{Re}_{\mathrm{Gülich}}^{z}, \tag{24}$$

Monte Verde [15] also correlated a relation to calculating the efficiency correction factor, similar to the previous three empirical models. Lastly, Ofuchi et al. [16] proposed an empirical correlation between Amaral et al. [7] experimental data and the findings of Ofuchi et al. [20, 21] numerical simulations. The model is based on the dimensional Ofuchi Reynolds number ($\mathrm{Re}_{\mathrm{Ofuchi}}$), defined as:

$$\mathrm{Re}_{\mathrm{Ofuchi}} = \frac{\omega_{\upsilon} Q_{w,\mathrm{BEP}}}{\nu\sqrt{gH_{w,\mathrm{BEP}}}} \frac{1}{\omega_{s,\mathrm{op}}}, \tag{25}$$

where ω_{υ} is the rotational speed under viscous flow, while $\omega_{s,\mathrm{op}}$ is a dimensional specific speed under operation [11], provided as:

$$\omega_{s,\mathrm{op}} = \frac{\omega Q_{\mathrm{op}}^{1/2}}{\left(gH_{\mathrm{op}}\right)^{3/4}}, \tag{26}$$

The author [11] states that in the above equation the rotational speed ω is the same rotational speed that results in $H_{w,\mathrm{BEP}}$ and $Q_{w,\mathrm{BEP}}$ in Eq. (26), and the op subscript stand for every operational condition retrieved from the water-based curves, which will be related to operation in viscous regime.

Now for the Ofuchi et al. [21] empirical model, the head correction factor is a function of the $\mathrm{Re}_{\mathrm{Ofuchi}}$ and the parameter x, where it is defined as:

$$x = -\left(\frac{b}{\mathrm{Re}_{\mathrm{Ofuchi}}^{c}}\right), \tag{27}$$

where b and c are empirical constants as shown in Table 1. Thus, the head correction factor C_H is equal to:

$$C_H = \mathrm{Re}_{\mathrm{Ofuchi}}^{x}, \tag{28}$$

Ofuchi et al. [20, 21] model made the same hypothesis of Stepanoff [8] to calculate the flow rate correction factor. Therefore, C_Q is calculated through Eq. (9). However, the Ofuchi model is similar to the Stepanoff model as it does not predict the lack of capability to calculate the efficiency correction factor [11].

These formulations offer simplified empirical tools that rely on minimal input parameters, such as rotational speed and viscosity, to estimate ESP performance under viscous flow regimes [11]. Nevertheless, the Monte Verde et al. [11] methodology was utilizing the original model parameter constants and then recording the optimized parameters results as shown below in Table 1 to compare error, mainly focusing on the BEP as the input for the error calculations.

In conclusion, correction factors are the core mathematical tools used in the empirical modeling of ESPs under non-water conditions. Each model discussed provides its framework for capturing viscosity-induced degradation in head and flow characteristics. These models serve as the analytical foundation for further parameter optimization and linearization, which are discussed in the following sections.

1.7.4 Characteristics of Nonlinear Empirical Models for Single-Phase Flow

Section 1.7.3 introduced correction factors based on empirical models, building on that this section outlines the main characteristics of previously mentioned nonlinear empirical models working under single-phase viscous fluid flow conditions [11]. While correction factors such as C_H, C_Q, and C_η help adjust water-based performance curves to account for viscous effects, their implementation and reliability vary by model [11]. Each model adopts a unique structure, some use viscosity dependent on Reynolds numbers, and others use specific speed relationships or fitted regression formulae as can be seen from the previous section [11]. Additionally, these models differ in terms of complexity, their applicability at or beyond the best efficiency point (BEP), and whether they account for flow-dependent variations in performance [11]. Table 2 summarizes and compares these models in terms of what they calculate, their assumptions, and the extent to which they adjust performance across changing flow conditions.

The evaluation of these empirical models for predicting ESP performance is done through MAPE, RMSE, and coefficient of determination (R^2), as mentioned by Monte Verde et al. [11] the Equations are defined as:

$$\text{MAPE} = \frac{1}{n}\sum_{i=1}^{n}\frac{|\hat{y}_i - y_i|}{y_i}.100, \tag{29}$$

Table 2 Comparative characteristics of nonlinear empirical models for single-phase flow [11]

Empirical model	Pump type	Operational range	Model complexity	Correction factors computed	Operating condition dependent
Stepanoff [8]	Volute centrifugal pumps	Up to 2020 mm^2/s; ω_S: 0.283–0.724	Low	C_H, C_Q	No (BEP only)
KSB [14]	Single-stage centrifugal	Up to 4000 mm^2/s; ω_S: 0.113–0.849	Medium	C_H, C_Q, C_η	Yes (CH only)
Gülich [13]	ESPs (general)	Up to 4000 mm^2/s; ω_S: 0.132–0.936	Medium	C_H, C_Q, C_η	Yes (CH only)
ANSI-HI [12]	Generic centrifugal pumps	Up to 4000 mm^2/s; $\omega_S < 1.097$	Medium	C_H, C_Q, C_η	Yes (CH only)
Monte Verde [15]	ESP GN5200	Up to 1069 mm^2/s; $600 < Re_{\omega_s} < 106$	High	C_H, C_Q, C_η	No (BEP only)
Ofuchi et al. [16]	ESP GN7000 + CFD models	Not explicitly stated; Operational ω_s calculated via $\omega_{S,OP}$	High	C_H, C_Q	Yes (CH only)

$$\text{RMSE} = \sqrt{\frac{1}{n}\sum_{i=1}^{n}(\hat{y}_i - y_i)^2}, \tag{30}$$

$$R^2 = 1 - \frac{\text{SSR}}{\text{TSS}}, \tag{31}$$

where y_i is the dataset that represents the experimental correction factors, $\hat{y}_i$ is the dataset of the calculated correction factors, and for n measured data $\overline{y}_i$ is the mean of the sample for that number of iterations. In Eq. 32, SSR is the sum of squared residuals and TSS is total sum of squares.

Thorough statistical comparison using large experimental dataset was conducted by Monte Verde et al. [11] by evaluating the prediction accuracy of these models. The Stepanoff [8]/TUALP model, originally limited by its iterative approach and BEP-only applicability, showed moderate performance, with unoptimized MAPE values for head and flow rate between 14.9 and 37.7% and RMSE reaching 0.203. After the empirical model parameter optimization, the model predictive accuracy improved, as it could achieve MAPE near 5.0%, RMSE of 0.042, and even a high value for R^2 reaching 0.935 [11]. The KSB [14] model, in its unoptimized state, had an MAPE of 22.1% for head and 24.9% for flow, with RMSE values of 0.106 and 0.119, respectively. Nevertheless, with as MAPE reached to 5.2% RMSE to 0.033, and R^2 as high as 0.980, it was concluded that the KSB model post-optimization has performed among the best [11].

In contrast, the Gülich [13] model exhibited relatively poor accuracy. Original MAPE values for head and flow reached 37.7% with RMSE around 0.174 and R^2 values even falling below zero in some cases, an indication of model mismatch [11]. Even after optimization, MAPE remained at 13.5% and RMSE at 0.094 [11]. The initially presented MAPE values were between 25.6 and 35.7%, with RMSE at 0.131 in the ANSI-HI [12] empirical model. Thus, similar trends were observed as the previous model were the optimized parameters showed an improvement to an MAPE of 11.5% and RMSE of 0.080, though this still trailed behind top-performing models [11].

The Monte Verde [15] own model steadily provided accurate predictions [11]. Originally, pre-optimization it had an MAPE of 18.1%, then post-optimization it fell to 5.9%. The same for RMSE going from 0.092 initially to 0.036, with R^2 reaching 0.977 after parameter optimization [11]. Finally, the Ofuchi et al. [16] model demonstrated strong baseline accuracy even before optimization. It achieved MAPE values between 2.8 and 7.5%, RMSE from 0.024 to 0.065, and R^2 values between 0.598 and 0.941, making it one of the most reliable models in head and flow estimation under varied operating conditions [11].

Collectively, these results reveal that not all empirical models perform equally when subjected to real operating data. Models such as KSB [14], Monte Verde [15], and Ofuchi et al. [16] stand out in terms of accuracy and adaptability, justifying their inclusion in this chapter.

1.8 Linearization Methods for Nonlinear Models

1.8.1 Importance and Methods of Linearizing Nonlinear Models

Linearization plays a pivotal role in simplifying nonlinear systems for easier analysis and control design. As explained by Cheng et al. [39], many real-world systems exhibit nonlinear behavior that complicates their direct analysis and modeling. Linearization transforms such systems into locally equivalent linear systems through coordinate transformations and state feedback techniques, which then allows the use of conventional linear system tools [39]. The transformation is often performed around an equilibrium point using the Jacobian matrix of the nonlinear system, resulting in a linear approximation of the original system dynamics. This transformation does not simplify the model itself globally but offers tractable representations for local behavior analysis near the operating point [39]. An example of such linearization is provided through the use of the Poincaré linearization theorem, which enables the conversion of a nonlinear system into a linear system, where is derived from the Jacobian of evaluated at the equilibrium point. This method ensures that the local dynamics of the nonlinear system can be analyzed with established linear techniques, such as those used in stability or controller design [39].

In a different but related context, Nguyen and Peraire [40] introduce the First-Order Empirical Interpolation Method (FOEIM), a technique designed for model reduction of nonlinear parameterized partial differential equations. Unlike the classical linearization through Jacobians, FOEIM focuses on efficient online evaluation of nonlinear terms by constructing reduced bases and selecting interpolation points that approximate the nonlinear function evaluations [40]. This reduction is particularly beneficial in iterative contexts such as real-time optimization or control, where multiple model evaluations are required. The key benefit emphasized by Nguyen and Peraire [40] is not linearization to be precise, but rather the computational efficiency achieved by decoupling offline expensive computations from online evaluations, leading to faster overall iterative performance in practice.

Thus, while [39] focus on mathematical linearization for enabling analysis of nonlinear control systems, Nguyen and Peraire [40] propose model reduction techniques to efficiently manage complex nonlinear computations during repeated simulations. Both approaches aim to handle nonlinear complexity, but their goals and methodologies differ: the former enhances theoretical tractability near equilibrium points, while the latter supports computational efficiency in large-scale parametric systems.

1.8.2 Optimization Techniques for Parameter Estimation

Parameter estimation in nonlinear system models relies heavily on optimization techniques that iteratively adjust model parameters to minimize discrepancies between predicted and observed values. This process is typically formulated as the minimization of objective functions, which represent statistical error metrics such as mean absolute error (MAE), RMSE, or MAPE [41].

Among the most widely adopted gradient-based methods is the Levenberg–Marquardt (LM) algorithm. Kumar [41] characterizes LM as a hybrid technique that interpolates between the Gauss–Newton and steepest descent methods through a damping parameter. This adaptability allows LM to perform reliably for well-behaved, smooth systems. However, the specific benefits attributed to the damping parameter such as balancing speed and stability are not explicitly stated by Georgios et al. [42], who acknowledge LM's use in load model identification but do not detail its convergence mechanics.

Kumar [41] further compares LM with derivative-free methods like the Nelder–Mead algorithm. Nelder–Mead, as detailed in Kumar's evaluation, is particularly useful in optimizing non-differentiable or noisy functions due to its reliance on geometric operations such as reflection, contraction, and expansion. It performed particularly well for chaotic and highly nonlinear systems, such as the Van der Pol oscillator [43]. Contrary to claims in earlier drafts, Kumar's results suggest that Nelder–Mead, rather than LM, consistently yielded the lowest RMSE in several benchmark systems. The notion that Nelder–Mead exhibited slower convergence than LM was not supported by direct evidence in the dataset presented in the provided excerpts.

Georgios et al. [42] contribute by introducing hybrid optimization strategies. While they do not discuss Nelder–Mead specifically, their work emphasizes combining global and local search methods. They propose frameworks where a global method identifies a promising region of the solution space, which is then refined by a local method such as LM. This combined strategy balances exploration and exploitation in the optimization landscape.

Complementing these algorithmic discussions, Dumre et al. [44] conducted a comparative study of error metrics for linear regression models, shedding light on how the choice of metric influences optimization outcomes. Their findings indicate that RMSE tends to amplify larger residuals, making it more sensitive to outliers, while MAE offers a balanced representation of average error magnitude. They caution against over-reliance on MAPE in cases where actual values approach zero due to its potential for instability.

These observations align with Kumar [41], who stresses the importance of selecting an error metric that aligns with the physical significance of the model output. For instance, if large deviations are of primary concern, RMSE would be preferred, otherwise, MAE or a combination of metrics may be more appropriate.

In summary, while LM remains a popular choice for smooth systems with reliable initial estimates, Kumar's evaluations demonstrate that Nelder–Mead may outperform LM in practice for nonlinear systems. Georgios et al. [42] advocate for hybrid methods that leverage the strengths of both global and local optimizers. Dumre et al. [44] reinforce that the effectiveness of optimization depends not only on the algorithm but also on the metric used to guide parameter updates.

1.8.3 Metrics for Evaluating Model Accuracy

The accuracy of the optimized models must be assessed following parameter estimation, which can be achieved through performance metrics that quantify prediction error. Two of the most widely used metrics in this context are the root mean squared error (RMSE) and mean absolute error (MAE). For instance, [45] state that RMSE due to the squaring of differences between observed and predicted values, this assessment procedure provides a measure of error magnitude and penalizes larger error more severely as a result. In contrast, MAE calculated as:

$$\text{MAE} = \frac{1}{n}\sum_{i=1}^{n}|y_i - \hat{y}_i|, \tag{32}$$

where y_i is the actual/experimental value, $\hat{y}_i$ is the predicted value, and n is the number of observations, offers a linear measure of the average absolute deviation, making it more interpretable in terms of average error size [45].

MAPE, or mean absolute percentage error, is another widely used metric, particularly for understanding relative errors in percentage terms. However, when the actual values come close to zero leading to disproportionately large error values, thus making it unstable. Due to the ease of interpretation of MAPE method [45], caution is usually advised to account for the method's high sensitivity.

In addition to these error metrics, the coefficient of determination (R^2) is a common statistical measure used to evaluate the explanatory power of a model. Chicco et al. [17] state that the dependent variable that is predictable from the independent variable, the R^2 quantifies this proportion of variance. For example, an R^2 value approaching zero or even reaching a negative value means poor or weak fit, while a value close to 1 illustrates a strong model fit. Thus, Chicco et al. [17] indicate that R^2 practicality should be evaluated across direct error metrics like RMSE and MAPE.

Model performance can be weighed up via a pattern of metrics as indicated by Chai and Draxler [45] and Chicco et al. [17]. The MAE error metric provides spontaneous average error, while MAPE contributes through a relative error perspective, and RMSE suitable is in penalizing large error deviations. However, R^2 proposes insight into the model's descriptive strength [17]. Together, these tools enable a comprehensive assessment of model validity.

Table 3 Representative experimental data at 1800 rpm and 120 cP viscosity showing a small set of all data

Set points	Speed (rpm)	Viscosity (cP)	Flowrate (bpd)	Head (ft)
1	1800	120	669.04	13.52
2	1800	120	705.1	13.28
3	1800	120	828.35	12.81
4	1800	120	850.3	12.29
5	1800	120	917.16	11.86

2 System Description and Methodology

2.1 Research Design and Methodological Approach

The study used data analysis to assess how ESPs work under different viscosity conditions. The process included choosing a model, estimating a correction factor, optimizing the parameters, and evaluating the model. All the computations were carried out using MATLAB R2023a with the experimental data similar to the ones presented in Table 3.

Experimental data were taken from a paper by Shi et al. [9], and water data at a single speed were found in the manual from the pump manufacturer [46]. Two nonlinear empirical models were chosen because they fit the data well. The initial parameters from previous studies were used to calculate the correction factors.

The Nelder–Mead Simplex and Levenberg–Marquardt methods were used to optimize the model parameters. The accuracy of the model was checked using MAE, MAPE, RMSE, R^2, and Absolute Relative Error (ARE), and all the calculations were done using the same units.

2.2 Data Source and Experimental Setup Overview

2.2.1 Source of Experimental Data

The data used in this study were taken from the work of Shi et al. [9], who studied how a 14-stage TE2700 radial-type ESP performed in viscous flow conditions.

The tests were carried out at four different rotational speeds: 1800, 2400, 3000, and 3500 rpm. For each speed, fluid viscosities ranging from 25 to 520 cP were used, and the corresponding pump head (ft) and flow rate (bpd) were measured. Resulting in a total of 530 operational points. The table above shows a small portion of how the data looks like.

The test setup was a closed loop with a torque sensor, viscometer, pressure transmitters and flow meters. The pump speed was adjusted by a variable-speed AC motor and the oil temperature was kept constant by a pipe-in-pipe cooling system to ensure the oil remained at the same viscosity. The ESP was designed to run at a specific speed of 1600 and included five-blade impellers and nine-vane diffusers. The results from this study were used to test and adjust the empirical models [9].

2.2.2 Water Data from Manufacturer Manual

Water performance data for the TE2700 ESP at 3500 rpm were obtained from the manufacturer's technical documentation provided by Wood Group ESP, Inc [46]. The performance curve (see Fig. 2) includes head, power, and pump-only efficiency as a function of flow rate, with the best efficiency point (BEP) located near 2700 barrels per day (BPD). The fluid used for this baseline performance characterization was water with a specific gravity (S.G) of 1.0.

Since the provided curve only represents performance at a single speed (3500 rpm), the affinity laws were applied to extrapolate water-based head versus flow curves at other rotational speeds. According to the affinity laws (Eq. 2), the head

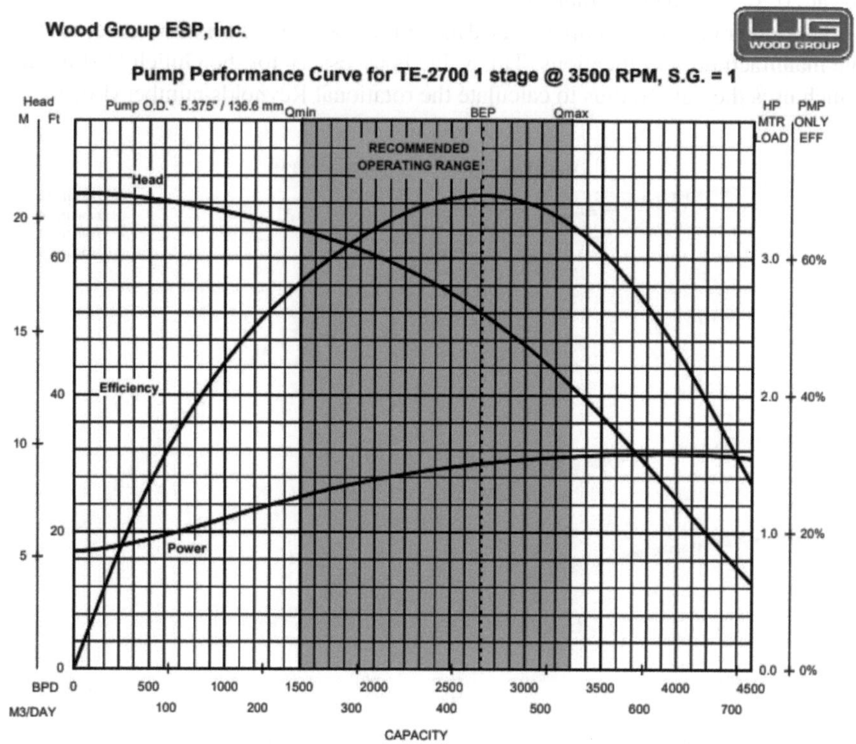

Fig. 2 Manufacturer water performance curve for TE-2700 ESP at 3500 rpm. Retrieved from Wood Group ESP, Inc [46]

and flow rate can be scaled with respect to rotational speed, assuming Newtonian, inviscid fluid flow and geometric similarity.

Using this approach, new curves were generated at 1800, 2400, and 3000 rpm, ensuring coverage across all speeds used in the experimental study. For each rpm, a total of 60 data points were computed to fully characterize the performance curve and allow accurate correction factor estimation at every operational range, thus in total we had a total of 240 water operational points across the rpm range.

2.2.3 Required Physical Parameters

In addition to flowrate and head, several supplementary parameters were required to identify the empirical models accurately and compute correction factors. These parameters include fluid density, kinematic viscosity, and the outer radius of the impeller, as well as the BEP conditions for water and the specific speed of the pump, shown in Fig. 3.

The working fluid used in the experiments was ISO VG 46 lubricating oil, which exhibits a kinematic viscosity of 46.98 mm^2/s and a density of approximately 860 kg/m^3 at 20 °C, as shown in Table 4.

The outer radius of the impeller used in the TE2700 ESP is 56.1 mm, retrieved from the manufacturer specifications. This value is necessary for the Gülich [13] model, which uses the outer radius to calculate the rotational Reynolds number (Eq. 15).

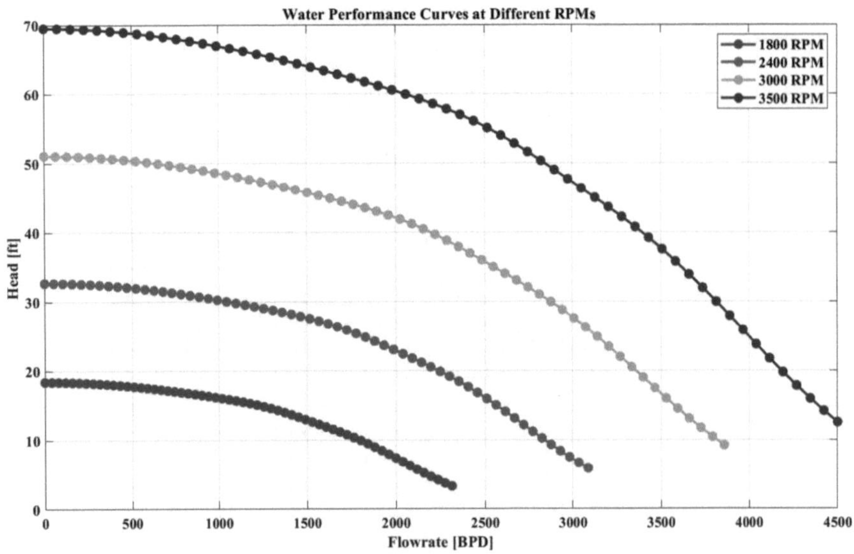

Fig. 3 Extrapolated water performance curves at 1800, 2400, 3000, and 3500 rpm using affinity law

Table 4 Properties of ISO VG lubricating oils

Viscosity grade ISO	Kinematic viscosity at 40 °C (mm^2/s)	Viscosity index (–)	Density at 20 °C (kg/m^3)
VG 22/B1	21.18	107	856.80
VG 32/B1	34.91	114	862.30
VG 46/B1	46.98	104	876.20
VG 46/B2	47.07	119	879.40
VG 68/B1	70.07	98	881.00
VG 100/B1	94.01	96	888.30

Retrieved from Biluš et al. [47]

Table 5 Calculated water BEP conditions and specific speed at various pump speeds

Speed (rpm)	Flowrate (BPD)	Head (ft)	Specific speed
3500	2700	52.17	0.5853
3000	2314.3	38.31	0.5854
2400	1851.4	24.54	0.5851
1800	1388.6	13.8	0.5785

The water-based BEP parameters at four different speeds were calculated using the affinity laws (Eq. 2). Flow rate and head at BEP for each speed were derived from the manufacturer's original 3500 rpm curve and scaled accordingly. Table 5 summarizes these values along with the corresponding calculated specific speed values. The specific speed (dimensionless form) was computed using Eq. (7).

All calculations were performed in SI units. The flow rate in BPD was first converted to m^3/s, and head from feet to meters. The resulting values of specific speed were confirmed to be consistent across all operating speeds, validating the similarity of the pump's operating behavior across different rotational speeds. The resulting values of specific speed were confirmed to be consistent across all operating speeds, validating the similarity of the pump's operating behavior across different RPMs.

2.3 *Selection of Empirical Models for Dataset Compatibility*

2.3.1 Justification for Selecting KSB and Gülich Models

The Gülich [13] and KSB [14] models were chosen because they are suitable for predicting the performance of electrical submersible pumps (ESPs) at different viscosity levels. While ANSI-HI [12], Monte Verde et al. [11], and Ofuchi et al. [16] only give correction factors at the best efficiency point (BEP), the KSB and Gülich models provide them for the entire range of operation.

The KSB model includes viscosity effects by using dimensionless parameters and adds a correction term (ξ) to represent deviations from BEP (Eq. 13). It also incorporates parameters that were either available in the dataset or could be estimated reliably, including specific speed and BEP conditions. The Gülich model employs the rotational Reynolds number (Eq. 15) to relate flow behavior with viscosity and impeller geometry. Its structure is well-suited for modeling high-speed, multistage ESPs and has been validated in prior literature.

These characteristics, along with data compatibility, made the KSB and Gülich models the most appropriate for parameter identification and full-range performance estimation in this study. These characteristics, along with data compatibility, made the KSB and Gülich models the most appropriate for parameter identification and full-range performance estimation in this study.

2.3.2 Attempted Linearization and Limitations

An initial effort was made to linearize the selected KSB and Gülich models to enable parameter estimation via multiple linear regression. However, the structural complexity of both models made linearization infeasible without distorting their physical meaning or violating their mathematical integrity.

The KSB model includes nested nonlinear relationships, such as logarithmic and exponential terms dependent on dimensionless variables and power-law exponents (e.g., Eqs. 11 and 13). These transformations involve interactions between multiple fitted parameters and variables, which could not be isolated into a linear form without introducing unacceptable simplifications.

Likewise, the Gülich model uses rotational Reynolds number (Eq. 15) and a viscosity-dependent exponent (Eq. 19) which lead to nonlinear expressions that cannot be separated linearly. When logarithmic transformations or approximations were used, the results did not fit well and lost their physical meaning. Because of these issues, the models were not changed and were estimated using optimization methods in their original nonlinear form. Given these limitations, both models were retained in their original nonlinear form for parameter estimation using optimization techniques.

2.4 Parameter Initialization and Correction Factor Estimation

2.4.1 Use of Original Parameters

To initiate the prediction of head and flow correction factors using the KSB and Gülich models, it was necessary to assign initial values to their empirical parameters. Rather than relying on arbitrary estimates, this study adopted the original empirical

Table 6 Original and optimized parameters used for KSB and Gülich models

Empirical model	Parameter	Original	Optimized
KSB [14]	a	15	10.41
	b	0.013	0.039
	c	0.165	0.113
	d	4	5.323
	e	0.25	0.390
	f	0.75	0.643
Gülich [13]	a	1.5	0.507
	b	6.7	12.09
	c	0.735	0.709

parameters developed by Monte Verde et al. [11], which were optimized using a database comprising six three-stage centrifugal ESP configurations namely P37, P47, P62, P100, HC10000, and HC12500.

These ESPs operated under four different speeds (1800, 2400, 3000, and 3500 rpm) and across a wide viscosity range (1–1273 mPa s), resulting in over 5800 experimental set points. The comprehensive nature of this dataset made the derived parameters a credible starting point, despite the difference in pump model compared to the TE2700 ESP used in the present study.

The original and optimized parameters selected for the KSB and Gülich models are summarized in Table 6. These served as the baseline values for subsequent correction factor prediction and model refinement.

2.4.2 Correction Factor Calculation Method

After acquiring the initial empirical parameters from literature (Sect. 2.4.1), all the necessary components were available to calculate the original correction factors for head (C_H) and flowrate (C_Q) using the KSB and Gülich models. The procedure was implemented in MATLAB R2023b and based on viscosity, RPM, and specific speed inputs.

To apply the models, several additional values were required: water performance data at BEP, specific speed for each rpm, kinematic viscosity (converted from dynamic viscosity using fluid density), and the outer impeller radius of the ESP. Water performance curves initially available only at 3500 rpm were extended to 3000, 2400, and 1800 rpm using the affinity laws (Eqs. 2 and 4).

To apply the models, additional inputs were needed: water performance data at BEP, specific speed per rpm, kinematic viscosity (from dynamic viscosity and fluid density), and the ESP impeller's outer radius. Water curves, initially available only at 3500 rpm, were extended to 3000, 2400, and 1800 rpm using the affinity laws (Eqs. 2 and 4), with 60 interpolated points per RPM to match the viscous dataset.

Table 7 Flow rate and head correction factors using the original parameters for both KSP and Gülich models at 1800 rpm

Speed (RPM)	Viscosity (cP)	C_Q_KSB	C_H_KSB	C_Q_Gülich	C_H_Gülich
1800	25	0.8329	0.9043	0.8856	0.8631
1800	120	0.6250	0.8626	0.7337	0.8275
1800	240	0.4810	0.7052	0.6294	0.6374
1800	520	0.3005	0.6765	0.4879	0.6358

Using these inputs and the original empirical parameters, correction factors were calculated and used to predict ESP performance under viscous flow. All original correction factors are included in the Appendix. The same procedure was repeated post-optimization, using updated parameters to improve accuracy; these optimized correction factors are also presented in Table 7.

Then the computed correction factors were applied to adjust the water performance data and predict ESP behavior under viscous conditions. For each RPM and viscosity, water head and flow values were scaled using KSB and Gülich correction factors. This was repeated across all data points, generating predicted curves for both models. These predictions were later compared to experimental results, with the resulting errors serving as a benchmark for parameter optimization.

2.5 *Model Prediction and Error Assessment*

2.5.1 Predicted Versus Experimental Data Comparison

After predicting the head and flow values using the original empirical parameters, the results were checked against the experimental data for every viscosity and RPM. It allowed us to see how much the actual results differed from the predictions.

The differences were then used to determine the initial error metrics which acted as a starting point for optimization. The errors in the baseline models helped to adjust the parameters, so the final models were more accurate in later stages.

2.5.2 Error Metric Formulae and Interpretation

To evaluate the accuracy of the predicted ESP performance, five error metrics were used: Mean absolute error (MAE), root mean square error (RMSE), mean absolute percentage error (MAPE), absolute relative error (ARE), and the coefficient of determination (R^2).

The MAE (Eq. 17) and RMSE (Eq. 18) provide measures of absolute deviation in feet, with RMSE giving greater weight to larger errors. Because MAPE (Eq. 19) and ARE in percentage terms, they help you see how well the model performs for different sizes of data. R^2 (Eq. 21) measures the amount of variation in the data that the model can explain. Since ARE formula was not previously introduced, it is expressed as:

$$\text{ARE} = \left| \frac{y_i - \widehat{y_i}}{y_i} \right| \times 100, \tag{33}$$

where y_i is the experimental value and $\widehat{y_i}$ is the predicted value. The results are given in percentage (%).

Looking at several metrics helps you see the whole picture. MAE and RMSE measure how much the model is off in total, while MAPE and ARE show how much the model is off about the data, which is useful for datasets of different sizes. R^2 also shows how well each model can predict the outcome.

2.6 *Parameter Optimization Procedure*

2.6.1 Optimization Algorithms Used

The best values for the empirical parameters in the KSB and Gülich models were found using the Nelder–Mead Simplex algorithm and Levenberg–Marquardt algorithm. Nelder–Mead was chosen because it does not depend on gradients and is suitable for searching in irregular areas, whereas Levenberg–Marquardt is good for solving nonlinear least-squares problems.

Both methods were used to compare how well they reduced the error in the model's predictions. Both methods were used to see which one worked best for the nonlinear models under different fluid conditions. The Levenberg–Marquardt optimization follows the parameter update rule:

$$\theta_{k+1} = \theta_k - \left(J^T J + \lambda I\right)^{-1} J^T r, \tag{34}$$

where θ_k is the current parameter vector, J is the Jacobian matrix of residuals, r is the residual vector $\left(y_i - \widehat{y_i}\right)$, λ is the damping factor, and I is the identity matrix. This formulation was implemented in MATLAB to minimize the sum of squared residuals between predicted and experimental head values. The Nelder–Mead algorithm, applied through MATLAB's *fminsearch* function, minimized the same error objective without requiring derivative input. The choice of both methods ensured methodological robustness across varying initial parameter sensitivities and error landscapes.

2.6.2 Optimization with Full Dataset

This section presents the optimization procedures used to minimize the prediction error in the KSB and Gülich models using the complete experimental dataset. Two separate optimization routines were implemented in MATLAB: one using the Nelder–Mead algorithm and the other using the Levenberg–Marquardt algorithm. Each method aims to minimize the root mean squared error (RMSE) between predicted and experimental head values by tuning the empirical parameters of both models. The corresponding workflow for each routine is shown in the flowcharts below (Fig. 4).

2.6.3 Optimization with Split Data Strategy

To ensure the generalizability of the models and avoid overfitting, a data splitting strategy was also adopted. The entire experimental dataset was randomly divided into two subsets: 85% for training and 15% for testing. This allowed the model parameters to be optimized using the training data, while their predictive performance was independently evaluated using the testing data. The detailed split is shown in Table 8.

Once the data was split, two separate optimization procedures were carried out—one using the Nelder–Mead (*fminsearch*) algorithm [48, 49] and another using the Levenberg–Marquardt (*lsqnonlin*) method [48, 49]. The trained parameters were applied to both subsets, and their predicted results were saved and evaluated using error metrics.

To clearly illustrate the sequence of operations followed for both optimization strategies, Fig. 5 presents the corresponding flowcharts. Each chart outlines the step-by-step execution logic from loading the datasets to saving predictions and computing error metrics.

2.6.4 Overview of MATLAB Implementation

All procedures were executed in MATLAB. For each of the four viscosities and four RPMs, 60 points were generated, resulting in 960 total prediction points per model version (original, Nelder, and Levenberg). Correction factors were first applied to water data to estimate head and flow under viscous conditions. These predictions were compared to experimental values by matching flow rates and evaluating head differences. Each iteration recalculated correction factors and predictions, forming a structured dataset for comprehensive performance comparison across all model configurations.

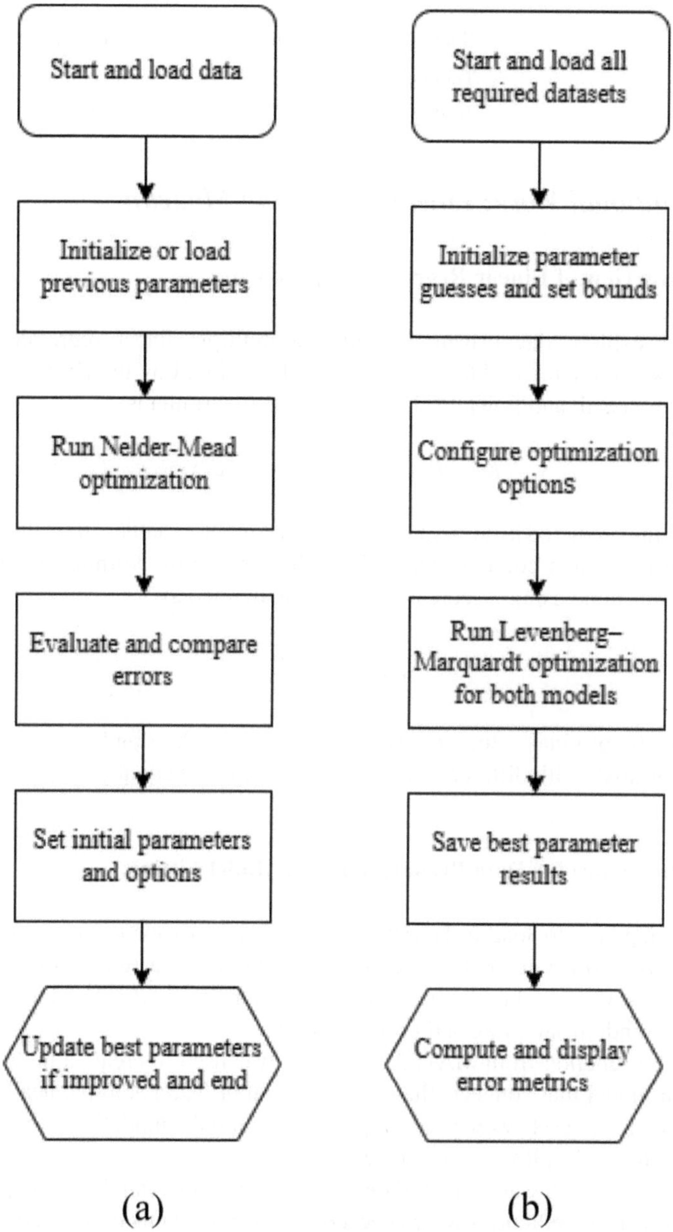

Fig. 4 Optimization workflow schemes for full-dataset parameter estimation. **a** Nelder–Mead optimization process. **b** Levenberg–Marquardt optimization process

Table 8 Train and split overview

Dataset	Percentage (%)	Number of points
Training	85	451
Testing	15	80

2.7 *Additional Work: Linear Regression Modeling*

2.7.1 Overview of Linear Regression Approach

To further explore data-driven modeling, a multiple linear regression (MLR) approach was implemented to predict the head generated by the ESP as a function of viscosity, speed, and flowrate. The general MLR formula is:

$$y = \beta_0 + \beta_1 x_1 + \beta_2 x_2 + \cdots + \beta_n x_n + \varepsilon, \quad (35)$$

where y is the predicted head, $x_1, x_2, \ldots, x_n$ are the input variables, β_0 is the intercept, β_n are the regression coefficients, and ε is the error term. Normalization of input variables was done using z-score normalization, defined as:

$$x_{\text{norm}} = \frac{x - \mu}{\sigma}, \quad (36)$$

where x is the original value, μ is the mean, and σ is the standard deviation. This ensured variables with different magnitudes were treated equally.

2.7.2 Procedure for Data Preparation and Model Fitting

The first step was to load a dataset that included experimental measurements of head, viscosity, flowrate, and rotational speed. The input features (viscosity, speed, and flowrate) were used to divide the dataset into training (70%) and testing (30%) subsets by random sampling and normalization.

The fitlm function from MATLAB [48, 49] was used to create a multiple linear regression model that connects the normalized input features to the head. Both the training and testing sets were used for predictions and the model's performance was checked using MAE, RMSE, R^2, and MAPE.

The trained model was then used on the entire experimental dataset to predict the head values. The predictions were checked against the results of actual experiments for every combination of viscosity and rotational speed. The comparison plots (Head versus Flowrate) were made to check how well the MLR model predicts in different operating conditions.

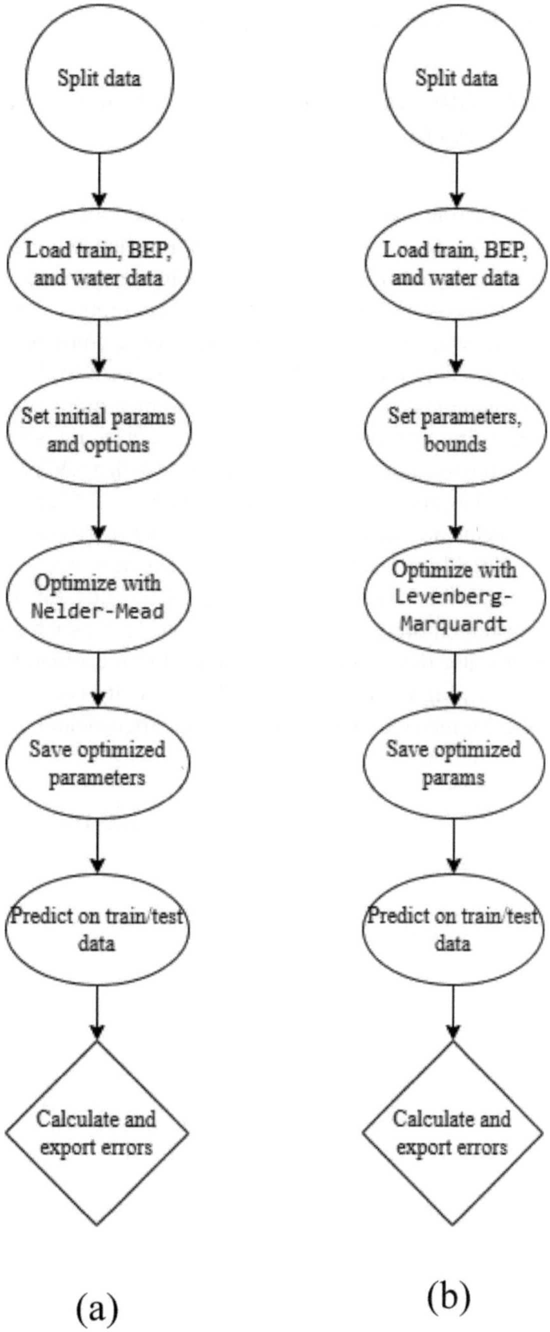

Fig. 5 Workflow of split data optimization using **a** Nelder–Mead and **b** Levenberg–Marquardt methods

3 Results and Discussion

3.1 *Introduction to Results: Impact of Viscosity on ESP Performance*

The section starts by looking at how fluid viscosity affects ESP performance, using water-based curves as a guide. The results from experiments at different viscosities and rotational speeds are compared to the water curve to find any differences in how the pump works. Figures 6 and 7 present representative cases for 1800 and 3500 rpm. As shown, the water performance curve consistently lies above the experimental curves. This indicates that higher viscosity leads to reduced head and flowrate at the same speed. The separation becomes more pronounced as viscosity increases, validating that viscosity significantly deteriorates ESP hydraulic performance.

The overall MAE reached 18.77 ft, with a RMSE of 20.60 ft. The MAPE averaged 259.31%, and the overall R^2 was negative -2.57, reflecting poor predictive alignment between the water model and viscous data across multiple viscosity-RPM conditions.

Notably, R^2 values become increasingly negative at higher viscosities, such as -73.78 for 520 cP at 1800 rpm, and -21.61 for 380 cP at 2400 rpm, indicating a substantial breakdown of affinity laws under such conditions. These results confirm that water-based curves are inadequate for estimating ESP behavior in viscous applications, highlighting the need for viscosity correction models. The complete table of error metrics comparing water to experimental data is included in Appendix 2 for full reference.

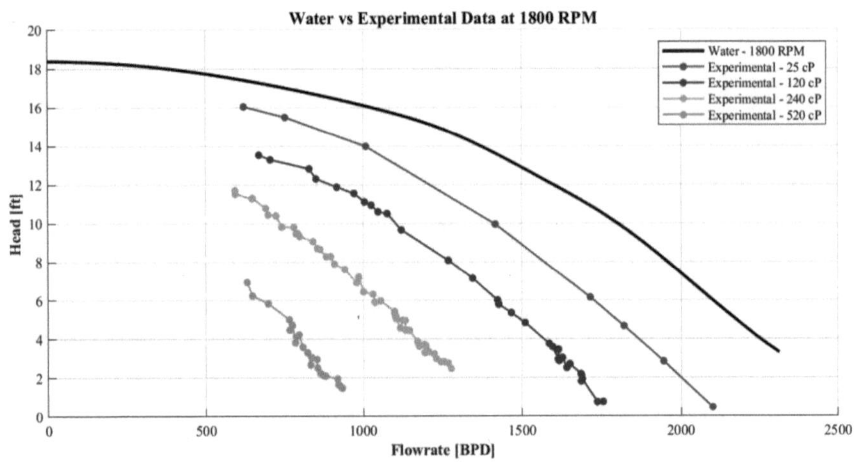

Fig. 6 Comparison of water versus experimental head to flow rate curve at 1800 rpm for various viscosities

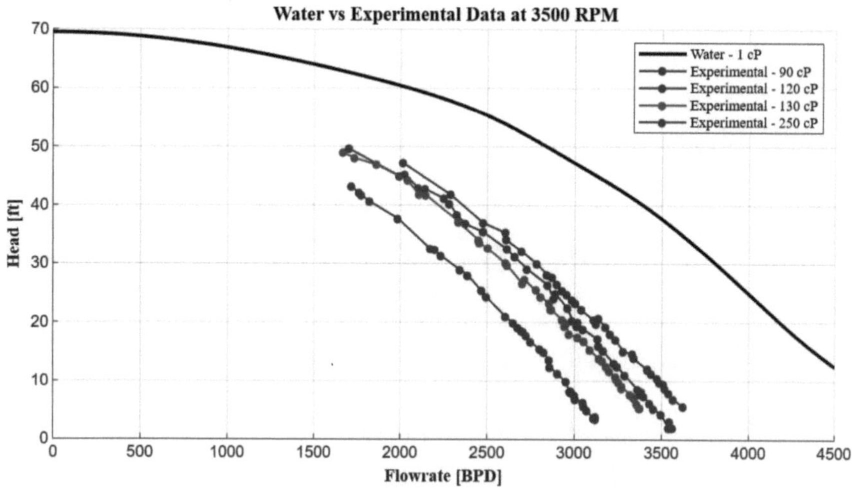

Fig. 7 Comparison of water versus experimental head to flow rate curve at 3500 rpm for various viscosities

3.2 Full Dataset Model Optimization (Without Splitting)

3.2.1 Evaluation of Predicted Performance Using Original Model Parameters

This section presents a comparison between predicted and experimental head values using the original, unoptimized parameters of the KSB and Gülich models. Figures 8 and 9 illustrate the residual head error (i.e., the difference between predicted and experimental head) as a function of flowrate for both models across all viscosity and rpm conditions. The KSB model (Fig. 8) shows a relatively balanced error distribution centered around zero, although some underprediction at lower flowrates is evident. In contrast, the Gülich model (Fig. 9) displays a systematic overprediction, with residual errors increasing consistently with flowrate.

As we can see from Fig. 10, performance curves for two representative high-viscosity cases 380 cP at 2400 rpm and 520 cP at 1800 rpm comparing the original KSB and Gülich model predictions to experimental results. The KSB model performs better than the Gülich model in tracking the head degradation due to viscosity, although both models deviate at higher flowrates.

The residual plots further emphasize the challenges in using unadjusted parameters for high-viscosity fluids. The KSB model, although slightly underpredicting at low flowrates, maintains tighter clustering around the zero-error line, indicating a more consistent performance across different flow and viscosity ranges. Meanwhile, the Gülich model tends to increasingly overestimate head values with rising flowrate, especially under high-viscosity conditions, pointing to the need for parameter recalibration.

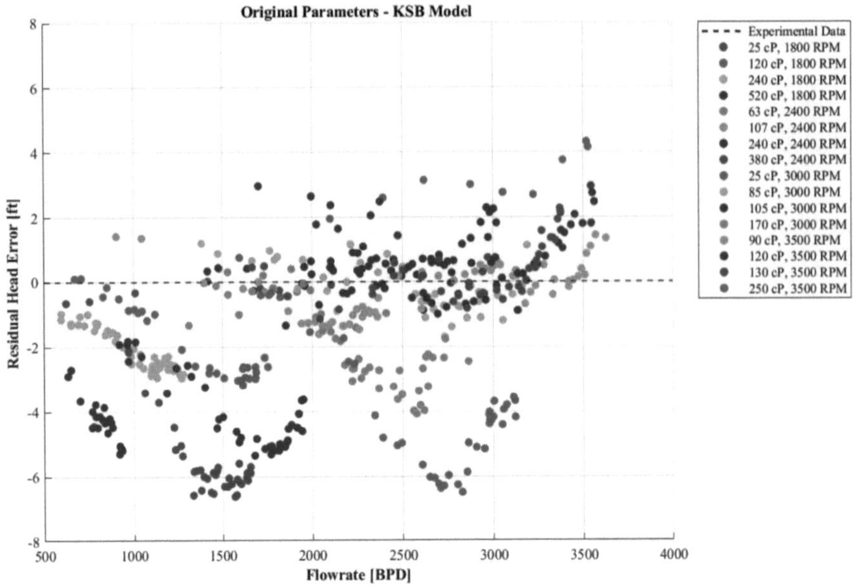

Fig. 8 Residual head error versus flowrate for KSB model using original parameters

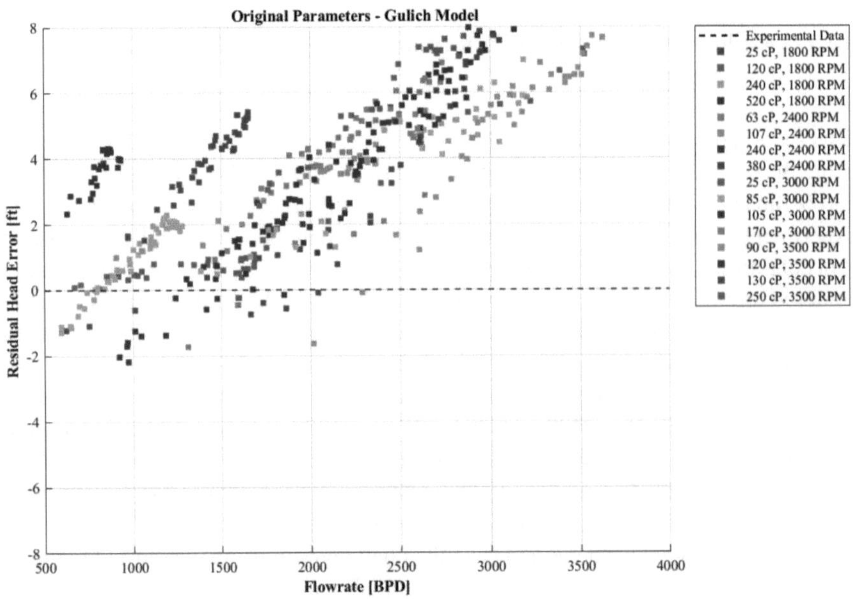

Fig. 9 Residual head error versus flow rate for Gülich model using original parameters

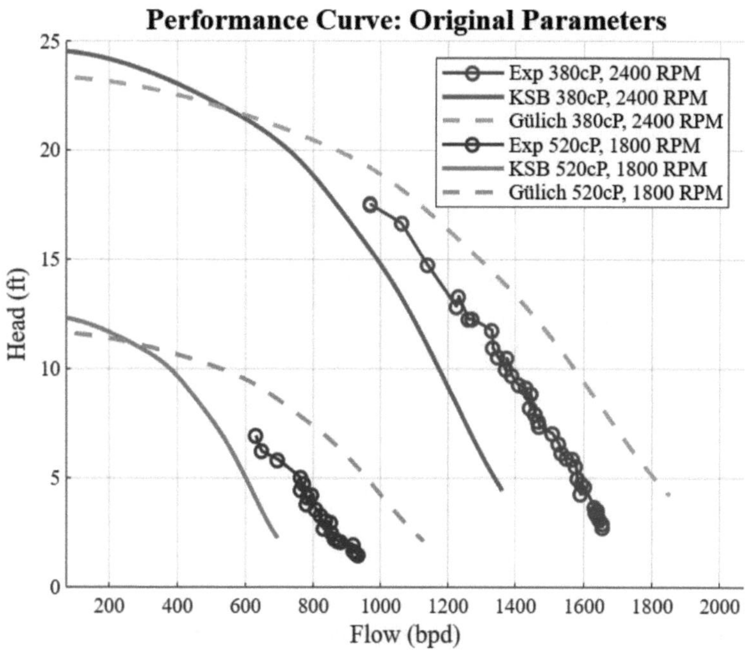

Fig. 10 Performance curve comparison for 380 cP at 2400 rpm and 520 cP at 1800 rpm: experimental versus predicted head using original parameters

3.2.2 Results After Nelder–Mead Optimization

The residual head error plot for the KSB model (Fig. 11) shows a significant tightening of the error band compared to its original form (Fig. 8). Previously, many data points, especially for higher viscosities like 380 and 520 cP, exhibited underpredictions exceeding −6 ft. After optimization, these errors are mostly constrained within ±2 ft across the full flowrate spectrum. Notably, the error pattern present before has flattened out, especially in the low-to-mid flowrate range (1000–2500 BPD), indicating better parameter tuning and reduced model bias. The variance in residuals across viscosity and rpm groupings has also become more consistent, suggesting that the optimized parameters generalize better across diverse operating conditions.

In the Gülich model (Fig. 12), the residual distribution still displays a wider spread compared to KSB, although some improvements are visible when contrasted with Fig. 9. The strong positive error bias seen in the original residuals, particularly for low flowrates and low-viscosity cases (e.g., 25, 120 cP), has been reduced. However, residuals at high flowrates still exhibit increasing positive deviation, especially for cases like 380 cP at 2400 rpm and 250 cP at 3500 rpm. This suggests that although the Nelder–Mead method improved the fitting, the Gülich model's correction structure may not fully accommodate high-viscosity nonlinearities.

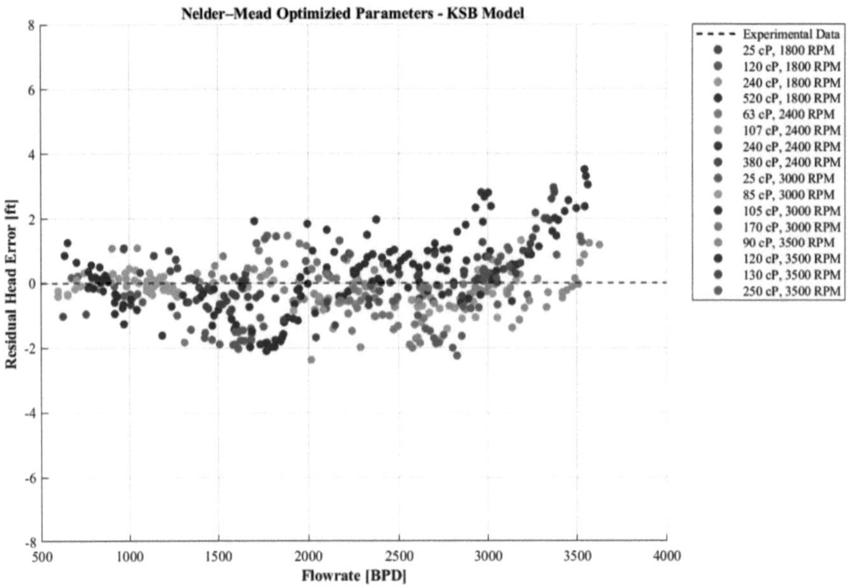

Fig. 11 Residual head error versus flowrate for Nelder–Mead optimized KSB model

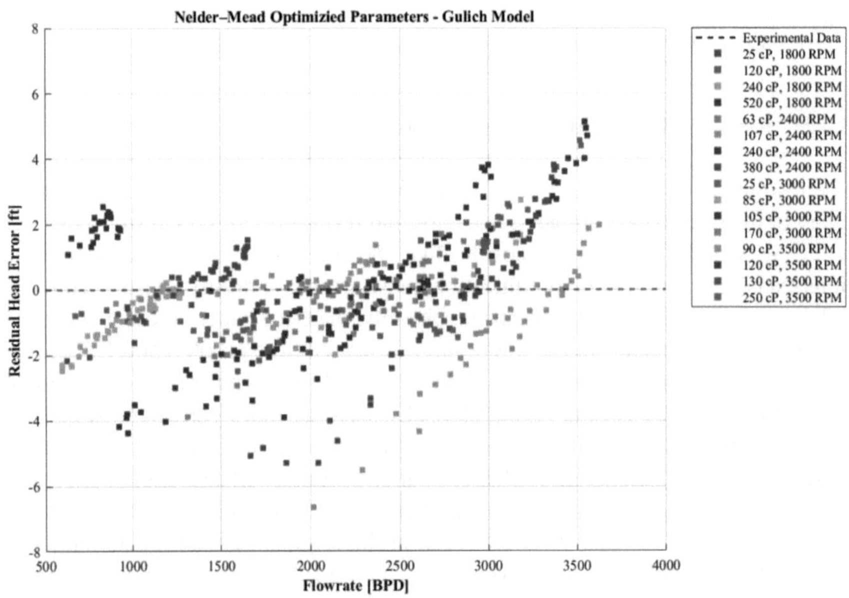

Fig. 12 Residual head error versus flowrate for Nelder–Mead optimized Gülich model

The head to flow rate curve in Fig. 13 clearly illustrate the improved alignment of the KSB model with experimental data. For 380 cP at 2400 rpm, the KSB curve overlays the experimental points almost exactly along the entire flow range, whereas the original version (Fig. 10) overpredicted head beyond 1400 BPD. In the 520 cP, 1800 rpm case, the optimized KSB curve shows a clear improvement by catching the steep drop in head at higher flowrates which was not seen before. On the other hand, the Gülich model does not accurately predict the head for high viscosity and does not replicate the sharp curve near BEP.

The KSB model also shows better matching of the slope. The changes along the curve, especially close to 1000 BPD, are well preserved, suggesting that the model's behavior as well as its magnitude were adjusted during optimization. The Gülich model is closer to reality, but it still shows that the slope transition zones lag, suggesting that it is less flexible in responding to changes in viscosity.

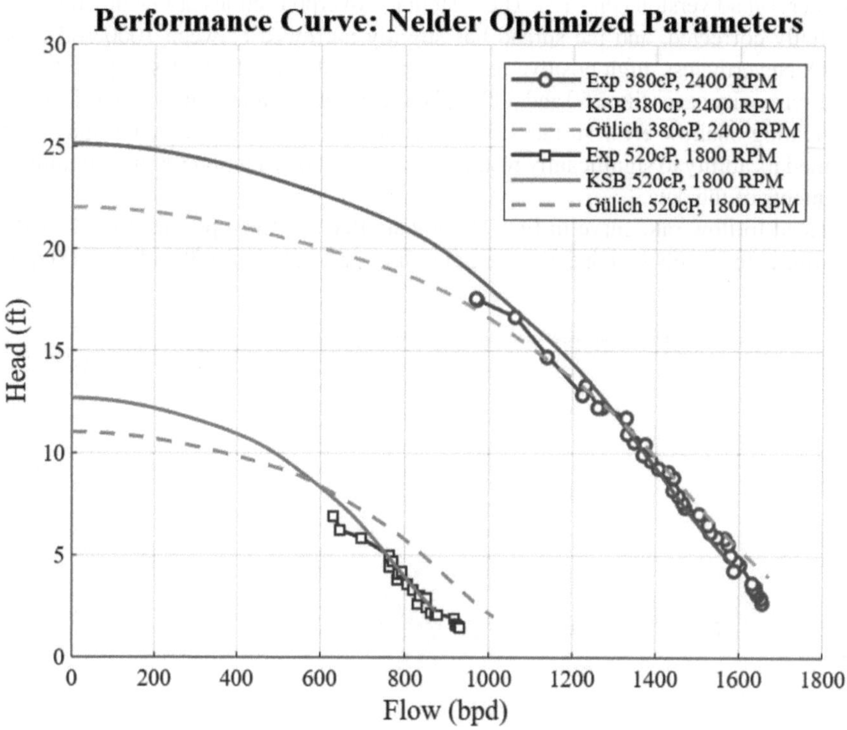

Fig. 13 Performance curve comparison for 380 cP at 2400 rpm and 520 cP at 1800 rpm: experimental versus predicted head using Nelder–Mead parameters

3.2.3 Results After Levenberg-Marguardt Method

Residual head error patterns for the KSB model following Levenberg–Marquardt optimization (Fig. 14) are remarkably similar to those seen under Nelder–Mead. The clustering around the zero-error line remains consistent across flowrates, and most points fall within ±2 ft, indicating effective parameter convergence. Compared to the previous residual plot for KSB (Fig. 11), the LM version shows slightly more symmetry in the central region (1500–2500 BPD) and a marginal reduction in over-estimation for high-viscosity datasets such as 380 and 520 cP. However, the tail-end deviations near 3500 BPD persist, suggesting that both optimization methods converge to similar local minima, with minor numerical differences in handling nonlinearity.

In the case of the Gülich model (Fig. 15), Levenberg–Marquardt provides a modest improvement in residual balance compared to its original (Fig. 9) and even to the Nelder–Mead version (Fig. 12). The systematic overpredictions at low flowrates are partially corrected, and the spread has slightly narrowed between 1000 and 2500 BPD. Nevertheless, high positive residuals persist at upper flowrates, especially for high-viscosity cases like 250 and 380 cP. While the algorithm is known for efficiently navigating shallow error surfaces, its dependency on gradient direction may have limited its ability to significantly reshape the residual distribution for more nonlinear segments of the data.

Head to flow rate curve in Fig. 16 indicate that the LM-optimized KSB model continues to closely match experimental trends, particularly for 380 cP at 2400 rpm.

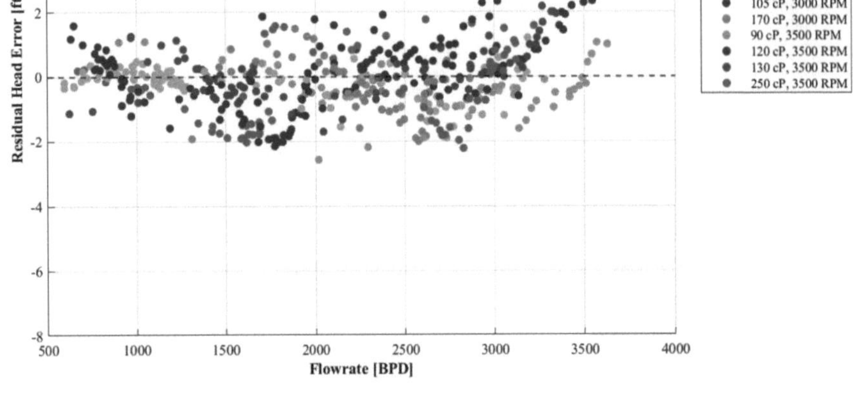

Fig. 14 Residual head error versus flowrate for Levenberg–Marquardt optimized KSB model

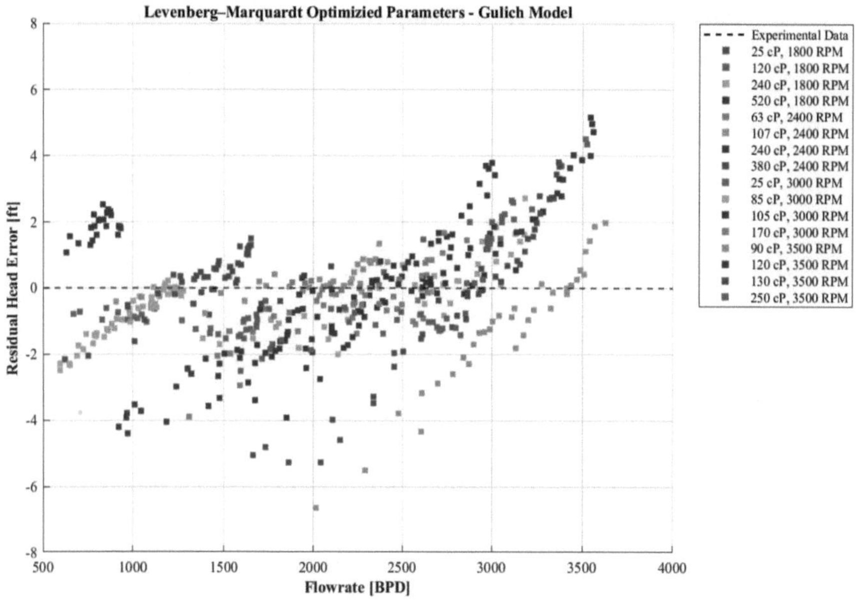

Fig. 15 Residual head error versus flowrate for Levenberg–Marquardt optimized Gülich model

The alignment is comparable to the previous curve in Fig. 13, affirming that LM achieved nearly the same accuracy as Nelder–Mead in replicating real pump behavior under this operating condition. For 520 cP at 1800 rpm, the KSB model again outperforms the Gülich model in capturing both head magnitude and curvature slope, though a slight underprediction remains in the highest flowrate zone. The Gülich model shows some curvature improvement relative to its original form but still fails to track the steep decline near the BEP region for 520 cP.

3.2.4 Quantitative Error Comparison Between Models and Methods

Although data splitting is a common practice to assess model generalization, the dataset used in this study was relatively small. Trial optimization using a 70/30 train-test split showed minimal differences from full-data model optimization. Therefore, as supported by Monte Verde et al. [11], the entire dataset was used to ensure comprehensive parameter fitting and more representative model behavior.

According to the MAE values, optimization greatly decreased the errors in both models. KSB had a MAE of 2.07 ft initially which was reduced to 0.749 ft by Nelder and 0.850 ft by Levenberg; Gülich's MAE went from 3.87 to 1.36 ft in both cases. MAPE went down a lot for KSB model and reached 13.2%, which is much lower than its original 42.5%. MAPE of Gülich model 25% after the optimization process. RMSE went down from 2.24 to 0.906 ft for KSB model and from 4.14 to ~1.64 ft for Gülich model (Fig. 17).

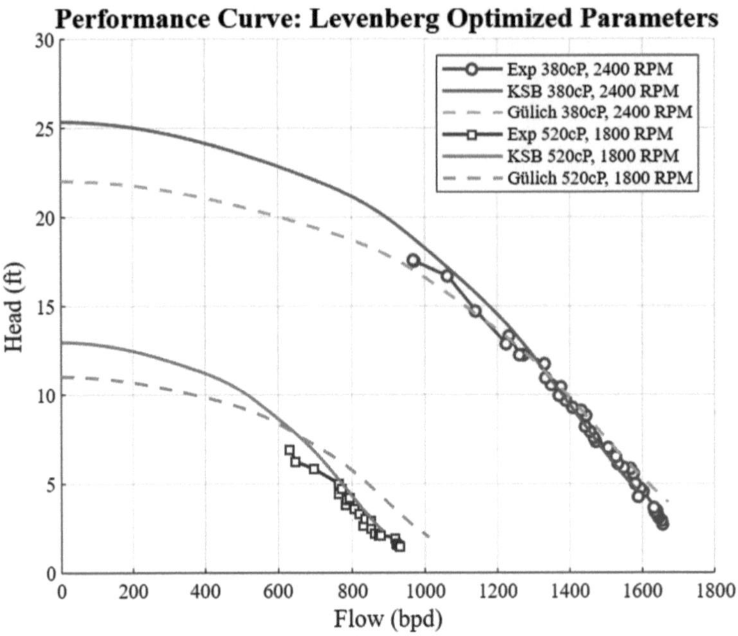

Fig. 16 Performance curve comparison for 380 cP at 2400 rpm and 520 cP at 1800 rpm: experimental versus predicted head using Nelder–Mead parameters

R^2 values confirm this pattern. The KSB model jumped from 0.286 to 0.977 (Nelder) and 0.970 (Levenberg), reflecting strong data fitting. Gülich improved from 0.399 to 0.857, but its performance was consistently lower than KSB.

Further insight is provided by examining absolute relative error across viscosity at specific speeds. At 1800 rpm (Fig. 18), both models showed large errors in their original form exceeding 160% for KSB and 130% for Gülich at 520 cP. After optimization, errors dropped below 20% across all viscosities, with the Levenberg-tuned KSB model performing best, especially at higher viscosities.

At 3500 rpm (Fig. 19), similar trends were observed. Original parameters for Gülich led to exceptionally high errors over 100% at 120 cP. KSB also exhibited moderate overprediction, but with lower magnitudes. Post-optimization, both models improved, with the KSB model under Levenberg optimization again achieving the lowest errors, staying below 10% across all viscosity levels.

These viscosity-specific results confirm that the optimized KSB model maintains higher accuracy and better consistency across the viscosity range. While both optimization methods reduced prediction errors, the Levenberg-tuned KSB parameters showed the strongest performance at low and high viscosities, especially under high-speed operating conditions. The optimized parameters table of optimization for the overall dataset is available in Appendix 2.

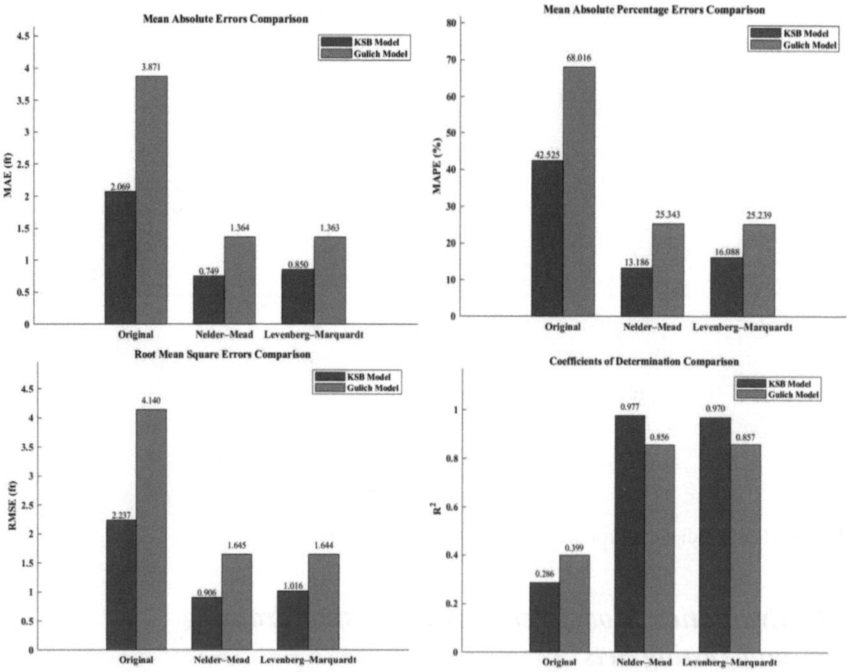

Fig. 17 Error metrics comparison between different parameter before and after optimization

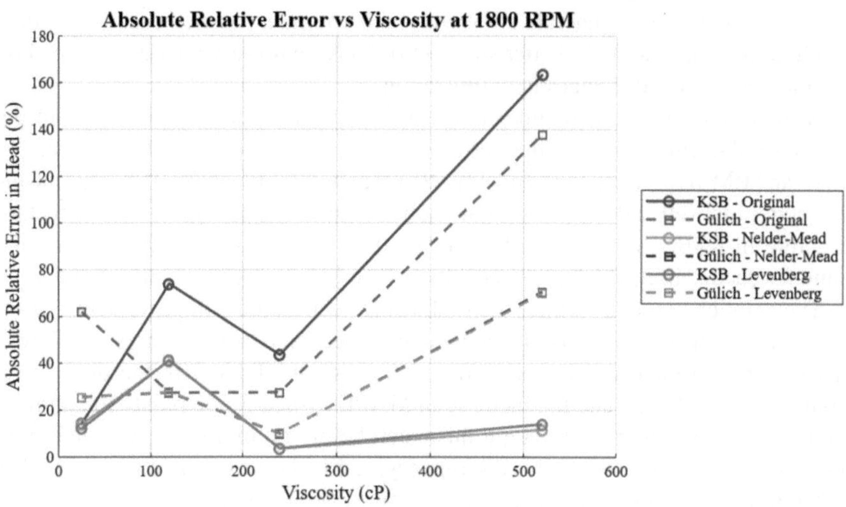

Fig. 18 Head prediction error across viscosities at 1800 rpm

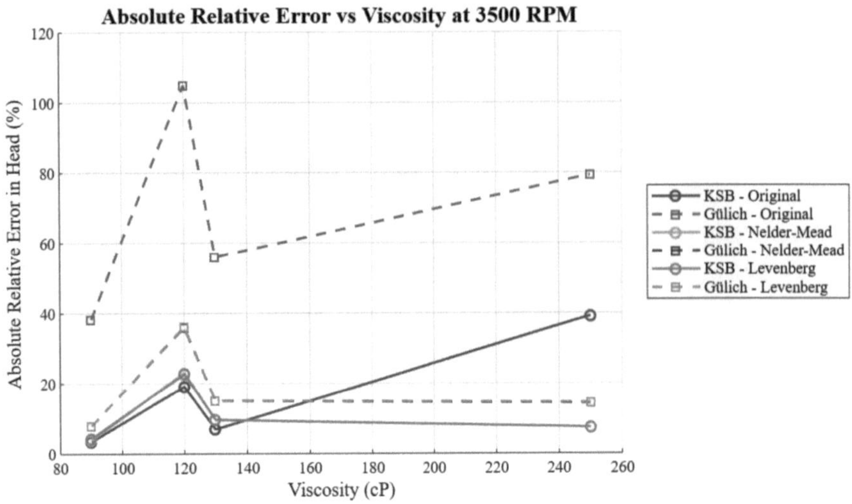

Fig. 19 Head prediction error across viscosities at 3500 rpm

3.3 *Evaluation Model Performance Using Training and Testing Sets*

By using an 85% training and 15% testing split, we can better see how each empirical model performs in generalization with both optimization methods. Rather than highlighting what should be better, this section looks at how each model and method react to new data and changes in optimization.

A main point in the KSB model is that the performance is similar in both the training and testing subsets. After optimization (as we can see in Table 9), the gap in MAE and RMSE between the two sets is <0.1 ft and R^2 is still above 0.98. This means the model fits well and can be used in many situations, showing that the correction structure in KSB is not affected by specific data points. The MAPE is almost the same for each subset which proves the model's reliability in different situations.

The Gülich model, while also showing substantial improvements post-optimization, displays slightly higher divergence between training and testing. Particularly in the original configuration (as we can see in Table 10), the drop in R^2 from 0.8587 to 0.7723 and the increase in MAPE by nearly 7% points signal more pronounced overfitting or structural mismatch before optimization. Although both optimization methods reduce this discrepancy, residual errors remain slightly higher on the test set than in KSB, especially in terms of MAE and RMSE.

The near-identical performance of Nelder–Mead and Levenberg–Marquardt in both models confirms that optimization technique does not drastically affect model generalization it is the underlying model structure that defines long-term predictive stability. This finding further supports the decision to adopt full-data optimization.

Table 9 Performance metrics of KSB model using 85% training and 15% testing split

Empirical model	Data set	Error metric	Parameter optimization method		
			Original parameters	Nelder–Mead simplex	Levenberg Marquardt
KSB [14]	Training (85%)	MAE [ft]	2.1578	0.8723	0.9250
		MAPE [%]	38.5206	12.9073	12.9073
		RMSE [ft]	2.8384	1.1228	1.1228
		R^2 [−]	0.9342	0.9897	0.9897
	Testing (15%)	MAE [ft]	2.3580	0.9294	0.9294
		MAPE [%]	34.3532	11.5029	11.5029
		RMSE [ft]	3.1995	1.2165	1.2165
		R^2 [−]	0.8948	0.9848	0.9848

Table 10 Performance metrics of Gülich model using 85% training and 15% testing split

Empirical model	Dataset	Error metric	Parameter optimization method		
			Original parameters	Nelder–Mead simplex	Levenberg Marquardt
Gülich [13]	Training (85%)	MAE [ft]	3.5163	1.2722	1.2726
		MAPE [%]	53.2744	18.7887	18.7749
		RMSE [ft]	4.1605	1.5909	1.5903
		R^2 [−]	0.8587	0.9793	0.9794
	Testing (15%)	MAE [ft]	4.1169	1.2942	1.2933
		MAPE [%]	60.1239	16.5334	16.5108
		RMSE [ft]	4.7073	1.6964	1.6940
		R^2 [−]	0.7723	0.9704	0.9705

3.4 Multi-linear Regression Results (Exploratory Analysis)

As part of an additional investigation beyond the empirical correction models, a multi-linear regression (MLR) approach was applied to predict head (H) as a function of flowrate (Q), viscosity (μ), and rotational speed (N). The final model was expressed as:

$$H = 14.42 - 5.0831\mu + 18.656N - 17.669Q, \tag{37}$$

This linear expression provided a good overall statistical fit, with $R^2 = 0.9367$ and an RMSE of 2.8229 ft, indicating that the model captured a significant portion of the variance in the dataset. However, while the global metrics suggest high accuracy, performance varied notably across different operating conditions.

As we can see from Fig. 20, the MLR model aligns well with experimental data for the 2400 rpm, 380 cP case, maintaining a close fit along the entire flow range. However, in the 1800 rpm, 520 cP case, the linear regression line fails to track the nonlinear decline in head, resulting in a prediction trend that diverges significantly from the experimental curve. This is consistent with the performance metrics, where the MAPE for 520 cP reached 153.06% in training and 221.21% in testing, and R^2 values turned negative reflecting model failure under extreme viscosity at low speeds.

Similarly, Fig. 21 shows the MLR fit at 3000 RPM for 105 cP and at 3500 RPM for 130 cP. In both cases, the predicted lines follow the experimental data more closely, especially at 105 cP, where the slope and curvature match well. This observation is supported by strong R^2 values above 0.96 and MAPE below 21% in both training and testing sets.

These results suggest that while the MLR model offers speed and simplicity, it struggles under highly nonlinear conditions such as high viscosity at low speed. The gap between training and testing errors, particularly the MAPE jump from 33.61 to 58.43%, confirms that this model lacks generalization capability across the full operational envelope. Nonetheless, MLR remains useful as a quick approximation tool when computational resources are limited or when high-fidelity models are unavailable.

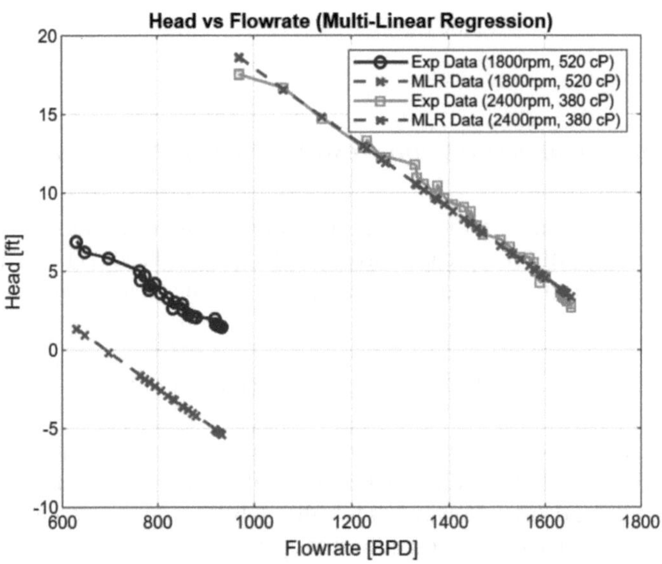

Fig. 20 MLR fit to experimental data at 1800 rpm (520 cP) and 2400 RPM (380 cP)

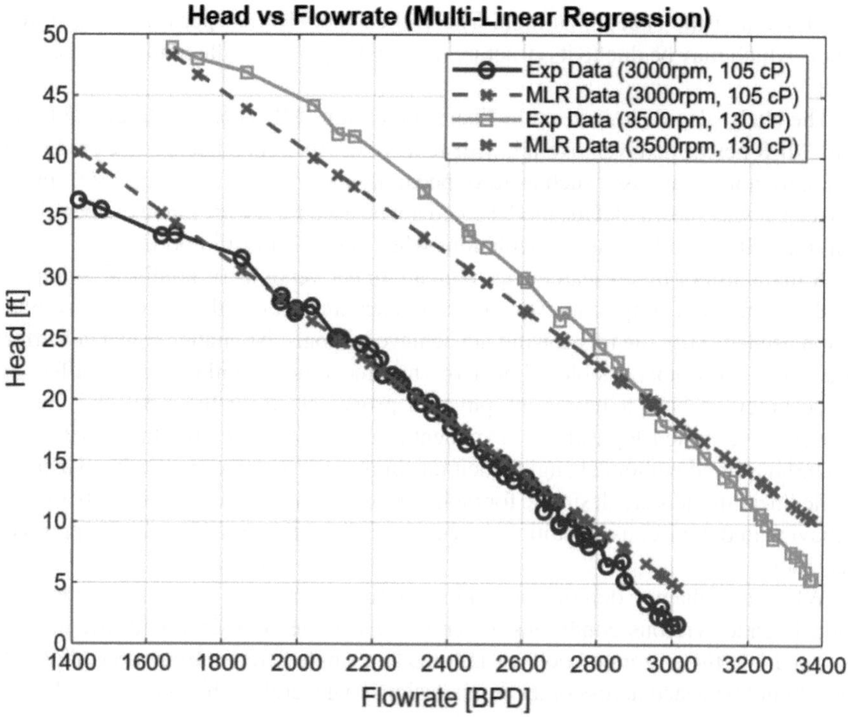

Fig. 21 MLR fit to experimental data at 3000 RPM (105 cP) and 3500 RPM (130 cP)

4 Conclusion

The reported research aimed to solve a main issue in ESP performance prediction when dealing with viscous fluids which is the absence of accurate, data-driven models that fit real-world situations. For a long time, the industry has relied on standard water-based performance curves, but this study reveals that they are not accurate in viscous flow regimes which can result in wrong pump choices, inefficient operations and higher expenses.

The study used experimental data from Shi et al. [9] gathered in a laboratory for a 14-stage TE2700 ESP and applied the Gülich [13] and KSB [14] models. They were chosen because they can correct for all types of errors which makes them useful for off-BEP analysis. The initial parameters from Monte Verde et al. [11] were used as the starting point and then they were improved using the Nelder–Mead Simplex and Levenberg–Marquardt algorithms.

It is evident from the results that using all the data leads to better generalization and a better performing model than splitting the data. The KSB model was especially strong, with little difference in error between the training and testing sets. Optimization helped reduce the MAE from 2.15 to 0.87 ft and the R^2 value increased to more than 0.98. Even after optimization, Gülich was less accurate, mainly when the fluid was thick and the speed was high. The Gülich model often performs worse than others which may be due to its structure not allowing it to handle nonlinear viscosity well.

The study also examined multi-linear regression (MLR) as a helpful additional tool. MLR performed well overall, with an R^2 of 0.9367, but it had trouble predicting in highly nonlinear cases such as low-speed, high-viscosity ones (e.g., 1800 rpm at 520 cP). It was shown that the model could not be used for a wide range of operations, but it is still useful for quick calculations when speed is most important.

A main aim of this research was to create linear versions of nonlinear empirical models to make it simpler to estimate parameters and reduce the amount of computation needed. Still, the mission did not achieve its goals. No matter how much effort was put in, it was not possible to linearize a nonlinear model without greatly affecting its accuracy or how it represents physical processes. Both the KSB and Gülich models have nested logarithmic, exponential and power-law structures that could not be simplified without changing their meaning. This underscores a critical insight: while linear models are desirable for their simplicity, the inherent complexity of ESP behavior under viscous conditions necessitates retaining nonlinearity to preserve accuracy.

While the chapter demonstrates meaningful progress in optimizing empirical models under viscous conditions, several opportunities remain for further investigation. First, the current models are tailored to a single ESP configuration (TE2700) and should be tested across other pump designs to generalize the findings. Including additional datasets covering higher and lower viscosities, as well as multiphase flow conditions, would enable more robust model training and validation.

Second, future research should incorporate the effects of gas–liquid interactions and solid content critical components in many field conditions but excluded here. Furthermore, exploring alternative empirical models or hybrid approaches that integrate physical laws with machine learning frameworks may provide more scalable and accurate solutions.

Third, since linearization failed for the models examined here, future studies could focus on local linearization techniques or piecewise linear approximations that retain physical meaning while allowing some degree of simplification. Alternatively, symbolic regression and surrogate modeling using AI techniques may offer new ways to extract tractable models from complex data.

In addition, using real-time ESP data with these models could help the models adjust to new conditions in the field. This would be in line with the oil and gas industry's efforts to use intelligent and autonomous systems for production.

In conclusion, the reported research contributes to empirical modeling of ESPs by thoroughly assessing correction factor-based models in viscous flow and creating a full model optimization framework. Although there are some challenges in linearization, the work offers important insights and tools that help improve ESP design by considering viscosity and it also points to future areas for improvement.

Appendices

Appendix 1: MATLAB Code Scripts

1. Nelder–Mead Optimization (*fminsearch*)

```
%% --- PARAMETER OPTIMIZATION USING NELDER-MEAD (fminsearch) ---
% Optimizes 9 empirical parameters (KSB + Gülich) to minimize
prediction error

% Initial guess: [KSB params (6) + Gülich params (3)]
theta0 = [10.41, 0.039, 0.113, 5.323, 0.390, 0.643, 0.507, 12.09,
0.709];

% Optimization options
options = optimset('Display','iter','MaxIter',2000,'MaxFunEvals',5000);

% Load required data
T = readtable('TogetCorrectionFactors.xlsx');
T.Properties.VariableNames =
matlab.lang.makeValidName(T.Properties.VariableNames);
expTrain = readtable('TrainSet.xlsx');
waterData =
readtable('Water_Combined_Head_vs_Flow_Data__All_RPMs_.csv');

% Run optimization
[theta_opt, ~] = fminsearch(@(theta) modelError(theta, T, expTrain,
waterData, 'Combined'), theta0, options);

% Display results

disp('Optimized Parameters (Nelder-Mead):');
disp(theta_opt);

%% --- MODEL ERROR FUNCTION ---
function [rmse] = modelError(theta, T, expData, waterData, modelType)
    % Extract parameters
    a = theta(1); b = theta(2); c = theta(3);
    d = theta(4); e = theta(5); f = theta(6);
    a_g = theta(7); b_g = theta(8); c_g = theta(9);

    % Constants
    r = 0.056; g = 9.81; bpd2m3s = 0.00000184013;

    all_true = []; all_pred = [];

    for i = 1:height(T)
        % Extract input values
        rpm = T.RPM(i); visc = T.Viscosity(i);
        omega_s = T.omega_s(i);
        Qw_BEP = T.Q_BEP_BPD(i) * bpd2m3s;
        Hw_BEP = T.H_BEP_ft(i) * 0.3048;
        Qw = T.Q_BPD(i) * bpd2m3s;
        nu = (visc / 0.898) * 1e-6;
        omega = 2 * pi * rpm / 60;
```

```
        % KSB Model
        B_HI = (480 * sqrt(nu)) / (Qw_BEP^0.25 * (g * Hw_BEP)^0.125);
        B = exp(log(B_HI) + 0.5 * (log(a) - log(52.933) -
log(omega_s)));
        CQ_KSB = exp(b * B * log(a / (52.933 * omega_s)) - c *
(log10(B))^d);
        xi = 1 - 0.014 * (B_HI - 1) * ((Qw / Qw_BEP) - 1);
        CH_KSB = (e + f * CQ_KSB) * xi;

        % Gülich Model
        Re_omega = (omega * r^2) / nu;
        Re_G = Re_omega * omega_s^a_g;
        x = -exp(log(b_g) - c_g * (log(Re_omega) + a_g *
log(omega_s)));
        CH_BEP = Re_G^x;
        CQ_G = CH_BEP;
        CH_G = 1 - (1 - CH_BEP) * (Qw / Qw_BEP)^0.75;

        % Get data subsets
        W = waterData(waterData.RPM == rpm, :);
        E = expData(expData.RPM == rpm & expData.Viscosity == visc, :);
        if isempty(W) || isempty(E), continue; end

        % Predictions
        if strcmp(modelType, 'KSB') || strcmp(modelType, 'Combined')
            Q_pred = W.Flow_BPD * CQ_KSB;
            H_pred = W.Head_ft * CH_KSB;
            H_interp = interp1(Q_pred, H_pred, E.Flow_BPD, 'linear',
'extrap');
            all_pred = [all_pred; H_interp];
            all_true = [all_true; E.Head_ft];
        end
        if strcmp(modelType, 'Gulich') || strcmp(modelType, 'Combined')
            Q_pred = W.Flow_BPD * CQ_G;

            H_pred = W.Head_ft * CH_G;
            H_interp = interp1(Q_pred, H_pred, E.Flow_BPD, 'linear',
'extrap');
            all_pred = [all_pred; H_interp];
            all_true = [all_true; E.Head_ft];
        end
    end

    % Compute RMSE
    err = all_true - all_pred;
    rmse = sqrt(mean(err.^2));
end
```

2. Levenberg–Marquardt Optimization (*lsqnonlin*)

```
%% --- PARAMETER OPTIMIZATION USING LEVENBERG-MARQUARDT (lsqnonlin) ---
% Optimizes 9 empirical parameters using nonlinear least squares

% Load datasets
T = readtable('TogetCorrectionFactors.xlsx', 'VariableNamingRule',
'modify');
T.Properties.VariableNames =
matlab.lang.makeValidName(T.Properties.VariableNames);
expTrain = readtable('TrainSet.xlsx');
waterData =
readtable('Water_Combined_Head_vs_Flow_Data__All_RPMs_.csv');

% Standardize variable names
expTrain.Properties.VariableNames{'Speed'} = 'RPM';
expTrain.Properties.VariableNames{'Flowrate'} = 'Flow_BPD';
expTrain.Properties.VariableNames{'Head'} = 'Head_ft';

% Initial parameter guesses
ksb_init    = [10.41, 0.039, 0.113, 5.323, 0.390, 0.643];
gulich_init = [0.507, 12.09, 0.709];
theta0 = [ksb_init, gulich_init];

% Bounds
lb = [1, 0, 0, 0, 0, 0, 0, 1, 0];
ub = [50, 1, 5, 10, 5, 5, 2, 20, 2];

% Options
opts = optimoptions('lsqnonlin', 'Display', 'iter', 'MaxIterations',
1000);

% Residual function
residualFun = @(theta) modelResiduals(theta, T, expTrain, waterData);

% Run optimization
[theta_opt, resnorm] = lsqnonlin(residualFun, theta0, lb, ub, opts);

% Display results
disp('Optimized Parameters (Levenberg-Marquardt):');
disp(theta_opt);

%% --- MODEL RESIDUAL FUNCTION ---
function residuals = modelResiduals(theta, T, expData, waterData)
    a = theta(1); b = theta(2); c = theta(3);
    d = theta(4); e = theta(5); f = theta(6);
    a_g = theta(7); b_g = theta(8); c_g = theta(9);

    r = 0.056; g = 9.81; bpd2m3s = 0.00000184013;
    residuals = [];
```

```
    for i = 1:height(T)
        rpm = T.RPM(i); visc = T.Viscosity(i); omega_s = T.omega_s(i);
        Qw_BEP = T.Q_BEP_BPD(i) * bpd2m3s;
        Hw_BEP = T.H_BEP_ft(i) * 0.3048;
        Qw = T.Q_BPD(i) * bpd2m3s;
        nu = (visc / 0.898) * 1e-6;
        omega = 2 * pi * rpm / 60;

        % KSB Correction
        B_HI = (480 * sqrt(nu)) / (Qw_BEP^0.25 * (g * Hw_BEP)^0.125);
        B = exp(log(B_HI) + 0.5 * (log(a) - log(52.933) -
log(omega_s)));
        CQ_KSB = exp(b * B * log(a / (52.933 * omega_s)) - c *
(log10(B))^d);
        xi = 1 - 0.014 * (B_HI - 1) * ((Qw / Qw_BEP) - 1);
        CH_KSB = (e + f * CQ_KSB) * xi;

        % Gülich Correction
        Re_omega = (omega * r^2) / nu;
        Re_G = Re_omega * omega_s^a_g;
        x = -exp(log(b_g) - c_g * (log(Re_omega) + a_g *
log(omega_s)));
        CH_BEP = Re_G^x;
        CQ_G = CH_BEP;
        CH_G = 1 - (1 - CH_BEP) * (Qw / Qw_BEP)^0.75;

        W = waterData(waterData.RPM == rpm, :);
        E = expData(expData.RPM == rpm & expData.Viscosity == visc, :);
        if isempty(W) || isempty(E), continue; end

        Q_KSB = W.Flow_BPD * CQ_KSB;
        H_KSB = W.Head_ft * CH_KSB;
        Q_G = W.Flow_BPD * CQ_G;
        H_G = W.Head_ft * CH_G;

        Hp_K = interp1(Q_KSB, H_KSB, E.Flow_BPD, 'linear', 'extrap');
        Hp_G = interp1(Q_G, H_G, E.Flow_BPD, 'linear', 'extrap');

        % Residuals: average of both models
        res = ((Hp_K + Hp_G) / 2) - E.Head_ft;
        residuals = [residuals; res];
    end
end
```

Appendix 2: Raw Data Tables and Plots

The list of values needed to get the correction factors at specific rpm, and viscosity.

Speed (rpm)	Viscosity (cP)	ω_s	$Q_{w,\mathrm{BEP}}$ (BPD)	$H_{w,\mathrm{BEP}}$ (ft)	Q_w (BPD)
1800	25	0.5853	1388.6	13.8	1763.314
1800	120	0.5853	1388.6	13.8	777.9429
1800	240	0.5853	1388.6	13.8	1348.457
1800	520	0.5853	1388.6	13.8	881.4857
2400	63	0.5854	1851.4	24.54	2350.971
2400	107	0.5854	1851.4	24.54	1037.143
2400	240	0.5854	1851.4	24.54	1797.943
2400	380	0.5854	1851.4	24.54	1175.657
3000	25	0.5851	2314.3	38.31	2938.629
3000	85	0.5851	2314.3	38.31	1296.686
3000	105	0.5851	2314.3	38.31	2247.429
3000	170	0.5851	2314.3	38.31	1469.486
3500	90	0.5785	2700	52.17	3428.571
3500	120	0.5785	2700	52.17	1512.686
3500	130	0.5785	2700	52.17	2621.829
3500	250	0.5785	2700	52.17	1714.286

Correction factor values retrieved using the original model parameters.

Speed (RPM)	Viscosity (cP)	C_Q_KSB	C_H_KSB	C_Q_Gülich	C_H_Gülich
1800	25	0.8329	0.9043	0.8856	0.8631
1800	120	0.6250	0.8626	0.7337	0.8275
1800	240	0.4810	0.7052	0.6294	0.6374
1800	520	0.3005	0.6765	0.4879	0.6358
2400	63	0.7676	0.8544	0.8367	0.8047
2400	107	0.6941	0.8964	0.7832	0.8596
2400	240	0.5442	0.7453	0.6756	0.6826
2400	380	0.4415	0.7527	0.6000	0.7154
3000	25	0.8711	0.9340	0.9148	0.8981
3000	85	0.7582	0.9267	0.8297	0.8897
3000	105	0.7300	0.8630	0.8092	0.8134
3000	170	0.6543	0.8658	0.7546	0.8255
3500	90	0.7707	0.8567	0.8380	0.8062
3500	120	0.7335	0.9153	0.8107	0.8774
3500	130	0.7221	0.8580	0.8025	0.8068
3500	250	0.6124	0.8441	0.7236	0.8034

Correction factor values retrieved using the Nelder–Mead optimized model parameters.

Speed (RPM)	Viscosity (cP)	C_Q_KSB	C_H_KSB	C_Q_Gülich	C_H_Gülich
1800	25	0.8294	0.8830	0.8446	0.8141
1800	120	0.6560	0.8554	0.6752	0.7897
1800	240	0.5472	0.7171	0.5700	0.5793
1800	520	0.3774	0.6903	0.4373	0.5998
2400	63	0.7693	0.8343	0.7874	0.7457
2400	107	0.7085	0.8802	0.7280	0.8239
2400	240	0.5961	0.7499	0.6157	0.6241
2400	380	0.5145	0.7693	0.5416	0.6739
3000	25	0.8379	0.8942	0.8804	0.8569
3000	85	0.7610	0.9051	0.7794	0.8572
3000	105	0.7372	0.8443	0.7564	0.7617
3000	170	0.6779	0.8547	0.6972	0.7846
3500	90	0.7707	0.8355	0.7867	0.7449
3500	120	0.7388	0.8943	0.7557	0.8418
3500	130	0.7295	0.8391	0.7465	0.7520
3500	250	0.6453	0.8379	0.6616	0.7593

Correction factor values retrieved using the Levenberg–Marquardt optimized model parameters.

Speed (RPM)	Viscosity (cP)	C_Q_KSB	C_H_KSB	C_Q_Gülich	C_H_Gülich
1800	25	0.8250	0.8770	0.8442	0.8136
1800	120	0.6556	0.8568	0.6750	0.7895
1800	240	0.5465	0.7211	0.5698	0.5791
1800	520	0.3827	0.7048	0.4369	0.5996
2400	63	0.7676	0.8316	0.7871	0.7453
2400	107	0.7083	0.8802	0.7277	0.8237
2400	240	0.5950	0.7522	0.6155	0.6239
2400	380	0.5148	0.7755	0.5414	0.6737
3000	25	0.8615	0.9059	0.8799	0.8564
3000	85	0.7597	0.9027	0.7791	0.8569
3000	105	0.7367	0.8432	0.7561	0.7614
3000	170	0.6777	0.8555	0.6969	0.7844
3500	90	0.7692	0.8329	0.7868	0.7449
3500	120	0.7386	0.8933	0.7559	0.8419
3500	130	0.7295	0.8386	0.7467	0.7522
3500	250	0.6452	0.8397	0.6619	0.7595

Error metrics between water data and experimental data head measurements.

Viscosity (cP)	Speed (RPM)	MAE (ft)	RMSE (ft)	MAPE (%)	R^2 (–)
25	1800	3.66	4.01	202.17	0.49
25	3000	12.14	12.54	348.00	0.05
63	2400	9.71	10.08	274.07	−0.63
85	3000	16.01	16.62	195.33	−1.24
90	3500	24.27	24.58	165.65	−4.60
105	3000	19.21	19.66	238.93	−3.63
107	2400	13.41	13.72	288.47	−3.62
120	1800	7.17	7.44	255.39	−2.10
120	3500	26.10	26.88	334.05	−2.50
130	3500	27.17	27.86	211.03	−3.69
170	3000	21.15	21.55	194.43	−5.74
240	1800	9.44	9.65	192.73	−10.68
240	2400	15.58	16.03	228.56	−5.92
250	3500	33.70	34.36	344.48	−7.05
380	2400	19.67	19.92	342.06	−21.61
520	1800	13.35	13.41	501.90	−73.78
Overall	Overall	18.77	20.60	259.31	−2.57

Parameter values after optimization with full dataset.

Parameter	Empirical parameter	Parameter optimization methods	
		Nelder–Mead simplex	Levenberg Marquardt
KSB [14]	a	0.125	1.270
	b	0.0776	0.0420
	c	5.466	0.631
	d	3.466	9.999
	e	0.337	0.355
	f	0.682	0.656
Gülich [13]	a	2.319	2.000
	b	6.357	6.762
	c	0.664	0.660

Parameter values after optimization with split dataset.

Parameter	Empirical parameter	Parameter optimization methods	
		Nelder–Mead simplex	Levenberg Marquardt
KSB [14]	a	0.195	1.002
	b	0.067	0.043

(continued)

(continued)

Parameter	Empirical parameter	Parameter optimization methods	
		Nelder–Mead simplex	Levenberg Marquardt
	c	102.893	1.392
	d	7.795	9.999
	e	0.292	0.357
	f	0.745	0.650
Gülich [13]	a	2.176	1.998
	b	6.481	6.708
	c	0.661	0.659

References

1. Mohammadzaheri M, Tafreshi R, Khan Z, Ghodsi M, Franchek M, Grigoriadis K (2020) Modelling of petroleum multiphase flow in electrical submersible pumps with shallow artificial neural networks. Ships Offshore Struct 15(2):174–183. https://doi.org/10.1080/17445302.2019.1605959
2. AL-Hejjaj MA, Sadeq DJ, Al-Fatlawi O (2023) A review of the electrical submersible pump development chronology. Iraqi J Chem Pet Eng 24(2):123–135. https://doi.org/10.31699/IJCPE.2023.2.14
3. Mohammadzaheri M, Tafreshi R, Khan Z, Franchek M, Grigoriadis K (2016) An intelligent approach to optimize multiphase subsea oil fields lifted by electrical submersible pumps. J Comput Sci 15:50–59. https://doi.org/10.1016/j.jocs.2015.10.009
4. Worth D, Al-Safran E, Choudhuri A, Al-Jasmi A (2019) Assessment of artificial lift methods for a heavy oil field in Kuwait. J Pet Sci Eng 180:835–843. https://doi.org/10.1016/j.petrol.2019.06.012
5. Fakher S, Khlaifat A, Hossain ME, Nameer H (2021) Rigorous review of electrical submersible pump failure mechanisms and their mitigation measures. J Pet Explor Prod Technol 11(10):3799–3814. https://doi.org/10.1007/s13202-021-01271-6
6. Gülich JF (1999) Pumping highly viscous fluids with centrifugal pumps: part 1. World Pumps 1999(395):30–34. https://doi.org/10.1016/S0262-1762(00)87528-8
7. Amaral G, Estevam V, Franca FA (2009) On the influence of viscosity on ESP performance. SPE Prod Oper 24(02):303–311. https://doi.org/10.2118/110661-PA
8. Stepanoff AJ (1949) How centrifugals perform when pumping viscous oils. Power. http://refhub.elsevier.com/S2949-8910(24)00241-0/sb18
9. Shi Y, Zhu J, Wang H, Zhu H, Zhang J, Zhang HQ (2021) Experiments and mechanistic modeling of viscosity effect on a multistage ESP performance under viscous fluid flow. Proceed Instit Mech Eng A J Power Energy 235(8):1976–1991. https://doi.org/10.1177/09576509211014974
10. Mohammadzaheri M, Ghodsi M, Ziaiefar H, Bahadur I, Zarog M, Emadi M, Soltani P, Amouzadeh (2023) A. Modelling of engineering systems with small data: a comparative study. In: Habib MK (ed) Perspectives and considerations on the evolution of smart systems. IGI Global, pp 120–136
11. Monte Verde W, Kindermann E, Bulgarelli NAV, Pastre LF, Foresti B, Bannwart AC (2024) A critical analysis and improvements of empirical models for predicting the performance of electrical submersible pumps under viscous flow. Geoenergy Sci Eng 238:871. https://doi.org/10.1016/j.geoen.2024.212871

12. ANSI-HI (2010) Rotodynamic pumps guideline for effects of liquid viscosity on performance. Hydraulic Institute
13. Gülich JF (2008) Centrifugal pumps. https://doi.org/10.1007/978-3-642-12824-0_1
14. KSB (1989) Kreiselpumpen-Lexiko. KSB. https://doi.org/10.1016/j.geoen.2024.212871
15. Monte Verde W (2016) Performance modeling of ESP pumps performance operating with gas-viscous liquid mixtures (Ph.D. thesis). University of Campinas [Universidade Estadual de Campinas]. https://doi.org/10.47749/T/UNICAMP.2016.976346
16. Ofuchi EM, Cubas JMC, Stel H, Dunaiski R, Vieira TS, Morales REM (2020) A new model to predict the head degradation of centrifugal pumps handling highly viscous flows. J Petrol Sci Eng 187:737. https://doi.org/10.1016/j.petrol.2019.106737
17. Chicco D, Warrens MJ, Jurman G (2021) The coefficient of determination R-squared is more informative than SMAPE, MAE, MAPE, MSE and RMSE in regression analysis evaluation. PeerJ Comput Sci 7:1–24. https://doi.org/10.7717/PEERJ-CS.623
18. Zhu J, Banjar H, Xia Z, Zhang HQ (2016) CFD simulation and experimental study of oil viscosity effect on multi-stage electrical submersible pump (ESP) performance. J Petrol Sci Eng 146:735–745. https://doi.org/10.1016/j.petrol.2016.07.033
19. Sirino T, Stel H, Morales REM (2013) Numerical study of the influence of viscosity on the performance of an electrical submersible pump. Am Soc Mech Eng Fluids Eng Div FEDSM 1:374. https://doi.org/10.1115/FEDSM2013-16374
20. Ofuchi EM, Stel H, Sirino T, Vieira TS, Ponce FJ, Chiva S, Morales REM (2017) Numerical investigation of the effect of viscosity in a multistage electric submersible pump. Eng Appl Comput Fluid Mech 11(1):258–272. https://doi.org/10.1080/19942060.2017.1279079
21. Ofuchi EM, Stel H, Vieira TS, Ponce FJ, Chiva S, Morales REM (2017) Study of the effect of viscosity on the head and flow rate degradation in different multistage electric submersible pumps using dimensional analysis. J Petrol Sci Eng 156:442–450. https://doi.org/10.1016/j.petrol.2017.06.024
22. Shojaeefard MH, Boyaghchi FA, Ehghaghi MB (2006) Experimental study and three-dimensional numerical flow simulation in a centrifugal pump when handling viscous fluids. IUST Int J Eng Sci 17(3):515
23. Li WG (2014) Mechanism for onset of sudden-rising head effect in centrifugal pump when handling viscous oils. J Fluids Eng Trans ASME 136(7):882. https://doi.org/10.1115/1.4026882
24. Stel H, Sirino T, Ponce FJ, Chiva S, Morales REM (2015) Numerical investigation of the flow in a multistage electric submersible pump. J Petrol Sci Eng 136:41–54. https://doi.org/10.1016/J.PETROL.2015.10.038
25. Syed FI, Alshamsi M, Dahaghi AK, Neghabhan S (2022) Artificial lift system optimization using machine learning applications. Petroleum 8(2):219–226. https://doi.org/10.1016/j.petlm.2020.08.003
26. Agwu OE, Alkouh A, Alatefi S, Azim RA, Ferhadi R (2024) Utilization of machine learning for the estimation of production rates in wells operated by electrical submersible pumps. J Petrol Explor Prod 14(5):1205–1233. https://doi.org/10.1007/s13202-024-01761-3
27. Zhu H, Zhu J, Zhang HQ (2022) Mechanistic modeling of gas effect on multi-stage electrical submersible pump (ESP) performance with experimental validation. Chem Eng Sci 252:288. https://doi.org/10.1016/j.ces.2021.117288
28. Kermit Brown (1980) The technology of artificial lift methods, p 2
29. Zhu H, Yu H, Sun Q, Wang Q, Jing H, Abdikadyrov R (2020) Deep learning-based performance prediction of electric submersible pumps under viscous and gas–liquid flow conditions. Machines 13(2):135. https://doi.org/10.3390/machines13020135
30. Mohammadzaheri M, Tafreshi R, Khan Z, Ziaiefar H, Ghodsi M, Franchek M, Grigoriadis K (2019) Modelling of electrical submersible pumps for petroleum multiphase fluids, an intelligent approach supported by a critical review and experimental results. J Eng Res 16(2):77
31. Abdalla R, Al-Hakimi W, Perozo N, Jaeger P (2023) Real-time liquid rate and water cut prediction from the electrical submersible pump sensors data using machine-learning algorithms. ACS Omega 8(14):12671–12692. https://doi.org/10.1021/ACSOMEGA.2C07609

32. Simpson AR, Marchi A (2013) Evaluating the approximation of the affinity laws and improving the efficiency estimate for variable speed pumps. J Hydr Eng 139(12):1314–1317. https://doi.org/10.1061/(asce)hy.1943-7900.0000776
33. Hydraulic Institute (1983) Hydraulic institute standards for centrifugal, rotary and reciprocating pumps. https://www.scribd.com/document/443342882/Hydraulic-Institue-Standard-14thed-pdf
34. Patil A, Morrison G (2019) Affinity law modified to predict the pump head performance for different viscosities using the Morrison number. J Fluids Eng Trans ASME 141(2):66. https://doi.org/10.1115/1.4041066
35. Sarbu I (1998) Energetic optimization of water pumping in distribution systems. https://www.researchgate.net/publication/259738612
36. Liu M, Tan L, Cao S (2020) Influence of viscosity on energy performance and flow field of a multiphase pump. Renew Energy 162:1151–1160. https://doi.org/10.1016/j.renene.2020.08.129
37. Cao GJ, Zhang WH, Chen J, Fu SH (2013) Performance correction chart of centrifugal oil pump for handling viscous liquids. IOP Conf Ser Mater Sci Eng 52:20. https://doi.org/10.1088/1757-899X/52/2/022020
38. Karassik IJ, Messina JP, Cooper P, Heald CC, York N, San C, Lisbon F, Madrid L, City M, New M, San D, Seoul J (2008) Fourth edition pump handbook
39. Cheng S et al (2010) Linearization of nonlinear systems
40. Nguyen NC, Peraire J (2023) Efficient and accurate nonlinear model reduction via first-order empirical interpolation. http://arxiv.org/abs/2305.00466
41. Kumar K (2023) Exploring optimization techniques for parameter estimation in nonlinear system modeling. http://arxiv.org/abs/2305.00351
42. Georgios B et al (2017) Optimization techniques for parameter estimation of dynamic load models. IEEE
43. Girotti M (2022) The Van der Pol oscillator
44. Dumre P, Bhattarai S, Shashikala HK (2024) Optimizing linear regression models: a comparative study of error metrics. In: Proceedings of the 4th international conference on technological advancements in computational sciences, ICTACS 2024, pp 1856–1861. https://doi.org/10.1109/ICTACS62700.2024.10840719
45. Chai T, Draxler RR (2014) Root mean square error (RMSE) or mean absolute error (MAE)? Arguments against avoiding RMSE in the literature. Geosci Model Develop 7(3):1247–1250. https://doi.org/10.5194/gmd-7-1247-2014
46. Wood Group (2009) Wood Group ESP, INC. USA Standard Terms and Conditions of Sale
47. Biluš I, Lešnik L, Kevorkijan L, Lovrec D (2021) Influence of differently viscous hydraulic fluid on the flow behaviour inside a hydraulic tank. In: International conference fluid power 2021: conference proceedings, pp 105–118. https://doi.org/10.18690/978-961-286-513-9.9
48. MathWorks (2025a) Fminsearch: search for local minimum of unconstrained multivariable function using derivative-free method—MATLAB. https://www.mathworks.com/help/matlab/ref/fminsearch.html#bvadxhn-13
49. MathWorks (2025b) Solve nonlinear least-squares (nonlinear data-fitting) problems, Matlab. https://www.mathworks.com/help/optim/ug/lsqnonlin.html

Publisher Correction to: Intelligent Control Strategy Optimization of Grid-Connected Hybrid PV-Battery-Supercapacitor Systems Using Machine Learning

Amer Bashir Omar Aboukhris and Mohammad Karimzadeh Kolamroudi

Publisher Correction to:
Chapter 1 in: A. Azizi (ed.), *Smart Diagnostics and Predictive Maintenance*, Emerging Trends in Mechatronics,
https://doi.org/10.1007/978-981-95-3921-5_1

The book was inadvertently published without incorporating the author corrections in chapter 1. The revised chapter is processed and the pagination is affected in the updated version.

The updated version of this chapter can be found at
https://doi.org/10.1007/978-981-95-3921-5_1

A. Azizi (ed.), *Smart Diagnostics and Predictive Maintenance*, Emerging Trends in Mechatronics, https://doi.org/10.1007/978-981-95-3921-5_7

MIX
Papier aus verantwortungsvollen Quellen
Paper from responsible sources
FSC® C105338
FSC
www.fsc.org

If you have any concerns about our products,
you can contact us on
ProductSafety@springernature.com

In case Publisher is established outside the EU,
the EU authorized representative is:
Springer Nature Customer Service Center GmbH
Europaplatz 3, 69115 Heidelberg, Germany

Printed by Libri Plureos GmbH
in Hamburg, Germany